Swordfish

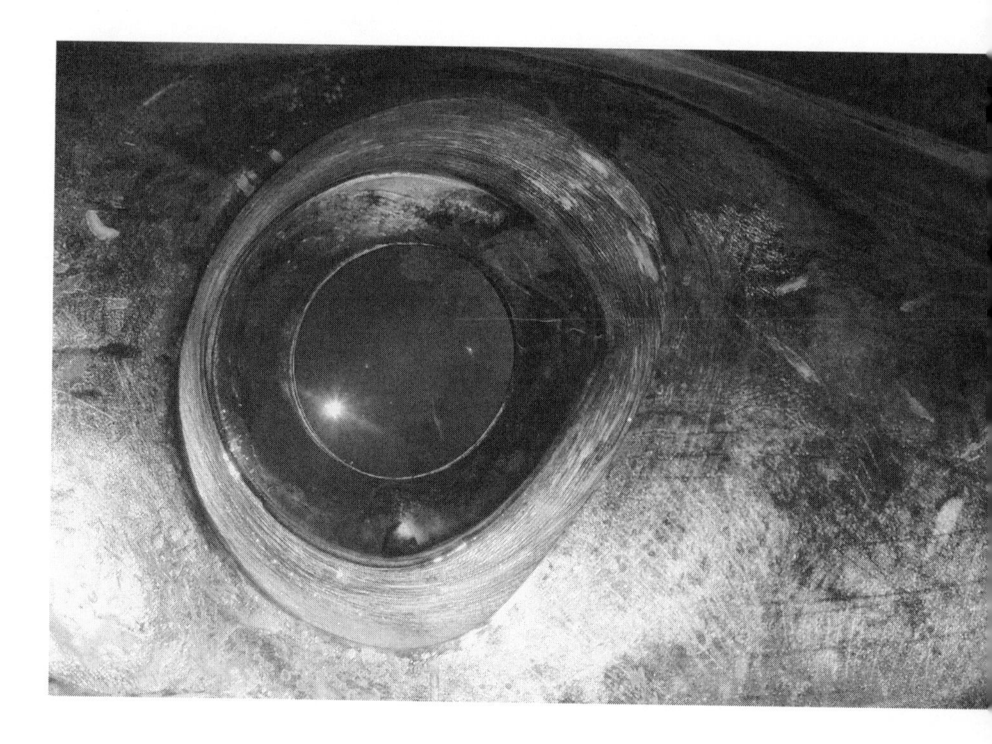

RICHARD ELLIS

Swordfish

A Biography of the Ocean Gladiator

THE UNIVERSITY OF CHICAGO PRESS

Chicago & London

Richard Ellis lives in New York and is the author of more than twenty books on marine life, including *Great White Shark*, *Men and Whales*, *Monsters of the Sea*, *The Encyclopedia of the Sea*, *Deep Atlantic*, *The Search for the Giant Squid*, *The Empty Ocean*, *Tuna: A Love Story*, *The Great Sperm Whale*, and *Shark: A Visual History*. A renowned painter of marine natural history, his paintings have been exhibited in museums and galleries around the world and have appeared in such publications as *Skin Diver*, *Audubon*, *National Wildlife*, *National Geographic*, and the *Encyclopedia Britannica*, as well as his own books.

The University of Chicago Press, Chicago 60637
The University of Chicago Press, Ltd., London

Printed in the United States of America
22 21 20 19 18 17 16 15 14 13   1 2 3 4 5
ISBN-13: 978-0-226-92290-4 (cloth)
ISBN-13: 978-0-226-92292-8 (e-book)
ISBN-10: 0-226-92290-1 (cloth)
ISBN-10: 0-226-92292-8 (e-book)

Library of Congress Cataloging-in-Publication Data

Ellis, Richard, 1938–
Swordfish : a biography of the ocean gladiator / Richard Ellis.
pages ; cm
Includes bibliographical references and index.
ISBN 978-0-226-92290-4 (cloth : alkaline paper) —
ISBN 978-0-226-92292-8 (e-book) 1. Swordfish. 2. Swordfish—Evolution. I. Title.
QL638.X5E45 2013
597'.78—dc23

2012034166

♾ This paper meets the requirements of ANSI/NISO Z39.48–1992 (Permanence of Paper).

FRONTISPIECE. *The eye of the swordfish.* Photograph by Carl Safina.

Contents

Preface

This is a single-species book, along the lines of *Great White Shark, The Search for the Giant Squid, Tuna: A Love Story, The Changing World of the Polar Bear*, and *The Great Sperm Whale.*

The shark book and the squid book had as their subjects marine creatures that threatened humans, either in fact or fantasy, and were thus perceived as having more popular appeal, than, say, a perch or a parrotfish. The worldwide population of bluefin tuna is threatened by overfishing, mostly to feed the insatiable Japanese sashimi market, and *Love Story* was written to draw attention the plight of the great and wonderful tuna. The story of the polar bear's relationship with humans is probably the most complex of all: we fear and love the great white bear, hunt it for sport, display it in zoos and circuses, and nominate it as the living symbol for global warming. And of course, the sperm whale's history is the most convoluted and contentious of all: the protagonist in America's greatest novel, as well as the protagonist in one of America's most important industries. I now bring you the broadbill swordfish, a creature that can hold its own against almost anyone and anything, including ships, boats, sharks, submarines, divers, and whales, but has not fared so well against the most effective predator the world has ever known, especially when he deploys those merciless engines of destruction known as longlines.

Among the authors who discuss big-game fishing, the billfishes usually have pride of place, but—until now—only the sailfish has its own book, Jim Bob Tinsley's *Swashbuckler of the Open Seas*. There are any number of books about salmon, and quite a few about the striped bass because it is the favorite of surf-casting fishermen and also the subject of one of the few successful conservation efforts in recent times. There are books about tuna, eels, herring, bluefish, and many of the lesser lights of the piscine fraternity. Although the swordfish has been the object of directed fisheries for centuries and has—at least according to some biologists—been close to extinction, I am delighted to conclude this introduction with an announcement that, as far as we know, *Xiphias gladius* has more or less successfully withstood centuries of harpooning, fishing, driftnetting, and longlining and has emerged bloody but unbowed. With help from members of the very species that threatened it, the world's swordfish populations are on the rise, and we may see *Xiphias gladius* resume its rightful position as one of the ocean's dominant predators, in character as well as numbers.

Like the tuna book, this, too, is a love story. How do we love the swordfish? As one of the most spectacularly beautiful animals on earth; as one of the largest and fastest, as well as the most heavily armed of all fishes; as the consummate pelagic predator, hunting just as efficiently in the chop of surface waters as in the unlit silences of the depths; as one of the ocean realm's most powerful hunters, fearing nothing from shark to man; as an attacker of animate and inanimate objects that would threaten its supremacy; as the sine qua non of big game fishes—the fighter against which all others are measured; as an animal that has steadfastly refused to reveal many of its deepwater secrets; as one of the world's favorite seafood items; as the object of some of the world's most intensive commercial fisheries . . . and as one of the large fish species that has been fished so heavily that its continued existence was once in question.

The swordfish is an apex predator, a fish that sits atop the food chain, preying on various creatures lower in the hierarchy. (The mako shark is also a rival of the swordfish and might even be considered its superior, as there are many more records of makos attacking swordfish than vice versa.) The reader will notice another large predator in this book's *dramatis animalae*, one that seems also to occupy an apex position but somehow finds itself one step below *Xiphias gladius*. The other apex predator is *Dosidicus gigas*, the jumbo or Humboldt squid. Swordfish, at a thousand pounds or more, are larger than these squid, which rarely reach two hundred pounds, but *Dosidicus* is considered one of the fastest, smartest, and most powerful of all cephalopods and might occasionally turn the tables. The stomach contents of swordfish have been extensively analyzed, but swordfish pieces in the stomachs of squid could only be identified by DNA analysis, and this has not been attempted very often.

I also encountered the vast body of lore and literature about *Dosidicus*, a squid that is as intriguing as its larger cousin, *Architeuthis*; probably more dangerous, and a whole lot better known. After all, the first ever photographs of a living *Architeuthis* were taken off Japan in 2004 and published in 2005, but *Dosidicus* has been known to Mexican and Peruvian fishermen for a long time and was photographed as long ago as 1940, in a *National Geographic* article about big-game fishing in the Humboldt Current. And while nobody fishes for *Architeuthis* (it reeks of ammonia and is inedible to humans), *Dosidicus* is the target of a major fishery in the Gulf of California. When I found myself researching *Dosidicus gigas* as a preferred prey item of *Xiphias gladius*, I realized that it was not only this interaction that was fascinating, it was also the two species, individually and together. Brad Seibel, a teuthologist (squid specialist) at the University of Rhode Island, suggested that a book about *Dosidicus* would sell better than a book about *Xiphias*, and he may be right. I chose the swordfish over the squid, but *Dosidicus* plays a big supporting role, and this may be as close as I'm ever going to come to writing a book about the Humboldt squid.

Many of those who helped in this project can be divided into two camps, the squid squad and the fish team. Among the squid researchers who helped me (and probably would have preferred to see a book about *Dosidicus*) are Freddy Arocha, Mike DeGruy, Bill Gilly, Angel Guerra, Roger Hanlon, Eric Hochberg, Christian Ibáñez, Unai Markaida, Ron O'Dor, Steve O'Shea, Uwe Piatkowski, Clyde Roper, and Brad Seibel. Mike DeGruy, who functioned in the fish and squid camps, died in a helicopter crash in Australia in February 2012. He would have liked this book.

The (sword) fish contingent includes Ed Beckwith, Peter Benchley, Charles Dana Gibson, Alessandro de Maddalena, Mark Ferrari, Harry Fierstine, John Graves, Silvestro Greco, Guy Harvey, Maria Laura Habegger, Bruce Knecht, Don Liberatore, Dick Lund, Brad Matsen, John McCosker, Stanley Meltzoff, Gail Morchower, Tom Moritz, Ted

Pietsch, Wes Pratt, Ed Pritchard, Mike Rivkin, Jay Rooker, Carl Safina, David Sanger, Mahmood Shivji, Mark Siddall, Arthur Spiess, Steven Stanley, Melanie Stiassny, Jan Timbrook, Ray Troll, and Boris Worm. Any errors that survived Harry Fierstine's readings and the comments by many people who know more than I do are attributable to my own obduracy or, as they used to say in the *New Yorker*, the fascination with the sound of my own words.

When I told my good friend Brad Matsen that I was planning to write a book about swordfish, he said he always wanted to write such a book, but in an incredible gesture of generosity, he turned over all of his notes to me. In his book *Deep Sea Fishing*, this is what he wrote about the swordfish:

> Off Rhode Island, the shimmer of morning above brings the big female swordfish to feed near the surface after a night in the depths. She works in the blackness on schools of mackerel and bluefish that are chasing feed downward in the vertical rhythm of light and dark. She moves with impossible grace and speed, the sea flowing over her smooth, scaleless body, perfectly shaped for what I like to think of as water-flying. Her cornflower blue eyes are enormous, the size of salad plates and best suited for the dim reflections of the abyss; the brightness of day is startling. She undulates constantly, transforming the power of her long muscles into motion and allowing her eyes to sweep a wide arc in search of food. She is pure hunter, an apex predator when mature, a meal only for humans and the largest sharks.

Maybe he should have written the swordfish book after all.

During the writing of this book and others—particularly *The Empty Ocean* and *Tuna: A Love Story*—I would often consult Ransom Myers about the facts and figures having to do with the massive worldwide depletion of marine fish species. I knew he was particularly interested in swordfish, and I was hoping he would read the manuscript for this book for what I expected would be an insightful and helpful review. Ram Myers died in 2007, and I never got to send him the manuscript. This would have been a much better book if he had had a chance to look it over.

1 Man Meets Swordfish

The earliest known inhabitants of what is now the state of Maine were the "Moorehead People," named for archaeologist Warren K. Moorehead, who excavated many of their sites. Four or five thousand years ago, these early "Native Americans" fished for cod and swordfish, probably from dugout or birchbark canoes. How do we know that? The canoes did not last, but archaeologists have examined shell mounds—known as middens, from the Danish word for trash heap—which contain the discarded shells of clams, oysters, mussels, and whelks and also traces of the culture that ate the mollusks or used them for bait. Buried among the shells are tools, including barbed hooks and harpoon heads made of deer bone. There are also pieces of the swordfish swords, suggesting that the Moorehead People hunted these great fish, and at that time, the swordfish were probably found a lot closer to shore than they are today. University of Maine archaeologist David Sanger has made a study of the middens and is convinced that swordfish formed an important part of the diet of these people, despite the fearsome reputation of the fish.

You probably wouldn't want to be in a birchbark canoe with an angry swordfish attached to it by a line, but Sanger discounts the pugnacious nature of swordfish strikes, claiming (as I do) that the fish couldn't determine that the source of its pain or discomfort was the canoe and that "attacks" on boats were mostly accidental. If the swordfish were sunning close to shore, the Moorehead harpooners would paddle out, heave the harpoon, lash the fish to the canoe, and butcher it at sea, bringing in only the most desirable pieces. Other archaeologists are not convinced of the "offshore butchering" hypothesis; Arthur Spiess and Robert Lewis found many more vertebrae and other bones at the Turner Farm site in Penboscot Bay and suggested that is was more likely that the fish were brought to shore to be cut up. Swordfish off the Maine Coast are not found close to shore nowadays, suggesting different conditions 4,000 years ago. Perhaps the water was warmer then, or there might have been other factors that drove the fish further offshore. The Moorehead People disappeared about 3,800 years ago, and the middens are topped with the refuse of people who evidently did not hunt swordfish, because no pieces of sword appear after that date. Did the Moorehead People die off because their food source moved away or were they overrun by another people?

On the opposite coast of North America, we find another aboriginal people intimately involved with swordfish. The Chumash of the Santa Barbara region of California were the finest boat builders among the California Indians. They made wooden plank canoes from redwood logs they obtained as driftwood from the Santa Barbara Channel. Their livelihood was based largely on the sea; they used over 100 kinds of fish and gathered clams, mussels, and abalone. But, as Davenport, Johnson and Timbrook wrote in their 1993 study, the evidence of a "special relationship" between the Chumash and the swordfish (which they called *elyewu'n*) "is available to us from a number of different sources: linguistic, ethnographic (recorded myths, ceremonial dances); archaeological (finds of swordfish parts, harpoon pieces, portrayals in Chumash art); and technical (fishing techniques and some facets of swordfish behavior)."

To the maritime Chumash, the swordfish was a creature with both material and ritual significance. They ate the meat, of course, but also used the sword and vertebral processes as spear points and digging implements; the large vertebrae were cut in half to make cups, and although it has not been preserved, there was a swordfish headdress made from the skull of the fish, thought to have been used in important ceremonial dances. The skull was decorated with shell inlay, especially around the large orbit, and there was a cape of iridescent abalone shells trailing behind the dancer, who was probably the village shaman. To the Chumash, swordfish was believed to be the chief of all the sea animals; the marine counterpart of human beings. When whales were stranded on the shore, they were said to have been driven ashore by swordfish to provide food for the people. The story of the swordfish attacking the whale is a part of many mythologies, from ancient Greek to California Indian. But even though the swordfish has no teeth and would have no reason to attack a whale, Davenport et al. include some of these stories as "eye-witness accounts."[1]

We will never know when the first fisherman cast his hook (or spear or net) into the water, but it is safe to assume that fishing is much older than civilization. The earliest fisherman caught their prey for subsistence; fish served as a plentiful source of protein. More or less contemporaneous with the Moorehead People, an Egyptian scene dating from around 2000 BC shows figures using a rod and line and also nets; in the ancient Minoan stronghold of Akrotiri on the Aegean island of Santorini, archaeologists have uncovered wall paintings that show two young men, each with strings of recently caught fish. The Minoan artists were so accurate that it is no problem to identify the fish they are holding: one man holds a string of bonito (*Sarda sarda*); the other has two strings of dolphinfish, *Coryphaena hippurus*. The island was partially destroyed in a giant volcanic explosion around 1600 BC, so the paintings are indisputable evidence of fishing more than 3,000 years ago. It is reasonable to assume that fishing was practiced wherever and whenever people believed there was seafood to be harvested.

Fish and cephalopods of all sorts were found in mosaics at Pompeii, buried by the ash from Mount Vesuvius in 79 AD Fishing is discussed by the chroniclers of ancient Greece and Rome, including Homer, Aristotle, Aelian, Pliny the Elder (who was killed by the eruption of Vesuvius), Strabo, and Pausanias. But, as William Radcliffe put it in his *Fishing from the Earliest Times*,

1. Unfortunately, one of the witnesses they relied on was Frank Bullen, an American seaman and author, whose 1898 *Cruise of the Cachalot* is cited as the source of some of the stories of swordfish attacks on whales. As we shall see, swordfish do indeed "attack" whales, but not the way Bullen described it. He relied more on his imagination than his experience for his accounts of the lives of swordfish in *Denizens of the Deep*, and it is now recognized that his work should be approached as imaginative fiction rather than descriptions of actual occurrences.

No character in Homer ever sailed for recreation or fished for sport. They were far too near the primitive life to find any joy in such pursuits. Men scarcely ever hunted or fished for mere pleasure. These occupations were not mere pastimes; they were counted on as hard labor. Hunting and fishing and laying snares for birds in Homer and even in the classical periods had but one aim: food.

Strabo, the Roman geographer who was born around 58 BC and died around 24 AD, traveled throughout the Roman world and recorded his observations in eight volumes. He described the manner in which they catch the swordfish off Sicily:

> One lookout directs the whole body of fishers, who are in a vast number of small boats, each furnished with two oars, and two men to each boat. One man rows, the other stands on the prow, spear in hand, while the lookout has to signal the appearance of a sword-fish. (This fish, when swimming, has about a third of its body above water.) As it passes the boat, the fisher darts the spear from his hand, and when this is withdrawn, it leaves the sharp point with which it is furnished sticking in the flesh of the fish: this point is barbed, and loosely fixed to the spear for the purpose; it has a long end fastened to it; this they pay out to the wounded fish, till it is exhausted with its struggling and endeavors at escape. Afterwards they trail it to the shore, or, unless it is too large and full-grown, haul it into the boat. . . . It sometimes happens that the rower is wounded, even through the boat, and such is the size of the sword with which the galeote (sword-fish) is armed; such is the strength of the fish, and the method of the capture, that in danger it is not surpassed by the chase of the wild boar.

Approximately a century after Strabo, the second-century AD Greek poet Oppian described fishing for swordfish in *Halieutica,* his hexameter poem on fishing:

> The fishermen fashion boats in the likeness of the Swordfishes themselves, with fishlike body and swords, and steer to meet the fish. The Swordfish shrinks not from the chase, believing that what he sees are not benched ships but other Swordfishes, the same race as himself, until the men encircle him on every side. Afterwards he perceives

his folly when pierced by the three-pronged spear; and he has no strength to escape for all his desire but perforce is overcome. Many a time as he fights, the valiant fish with his sword pierces in his turn right through the belly of the ship; and the fishers with blows of brazen axe swiftly strike all his sword from his jaws, and it remains fast in the ship's wound like a rivet, while the fish, orphaned of his strength, is hauled in. As when men devising a trick of war against their foes, being eager to come within their towers and city, strip the armour from the bodies of the slain and arm themselves therewith and rush nigh the gates; and the others fling open their gates as for their own townsmen in their haste, and have no joy of their friends; even so do boats in his own likeness deceive the Swordfish.

Moreover, when encircled in the crooked arms of the net the greatly stupid Swordfish perishes by his own folly. He leaps in his desire to escape but near at hand he is afraid of the plaited snare and shrinks back again; there is no weapon in his wits such as is set in his jaws, and like a coward he remains aghast till they hale him forth upon the beach, where with downward-sweeping blow of many spears men crush his head, and he perishes by a foolish doom.

Although Italian and Spanish fishermen have hunted *pesce spada* or *espadón* in the Mediterranean for centuries (and still do), the western North Atlantic swordfishery is a relatively recent development. Comparatively speaking, of course, the European settlement of the western shore of the North Atlantic is itself a recent development. The first published record of North Atlantic swordfish—and a swordfish attack on a boat—appears in John Josselyn's 1675 *Account of Two Voyages to New England*, in which he wrote, "In the afternoon [of June 20, 1638] we saw a great fish called the *vehuella* or Sword fish, having a long, strong, and sharp fin like a sword-blade on the top of its head, with which he pierced our Ship, and broke it off with striving to get loose, one of our sailors dived and brought it aboard."

Cod fishermen, first from Europe, and then from New England and Canada, had been working the offshore waters of the western North Atlantic from the time that John Cabot pulled up a basketful of cod in 1497, and it is unlikely that the fishermen overlooked those sickle fins that occasionally broke the surface. Regardless of its pugnacity, a big, firm-fleshed fish could not swim unmolested off New England for long, and by the early years of the nineteenth century, a swordfishery had begun. In David Storer's 1839 "Reports on the Fishes, Reptiles and Birds of Massachusetts," we find this description of the swordfish and the fishery:

> It is generally discovered by the projection of its dorsal fin above the surface of the water as it is pursuing shoals of mackerel upon which its feeds, about 15 or 20 miles from the shore of Martha's Vineyard. The fishermen capture it by means of an instrument called the "lily iron" from the form of its shafts or wings which resemble the leaves of a lily. The instrument is thrown like a harpoon with great force into the fish, the attempt always being made to wound the animal in front of the dorsal fin. . . . When unmolested, it not infrequently is observed to spring several times its length, several feet above the surface of the water.

In his 1887 history of the American swordfishery, G. B. Goode quotes a certain Captain Merchant of Gloucester, who told him that "the first swordfish ever brought to Gloucester within his recollection was caught on George's Bank around 1831 by Captain Pew who brought it in and sold it at the rate of $8 a barrel, salted." Before that, fishermen had been very much afraid of them, but afterward a good many were caught. Goode then identifies "the earliest record of its use for food . . . found in the *Barnstable Patriot* of June 30, 1841, in which it was stated that the fishermen of the island south of Cape Cod take a considerable number of these fish every year by harpooning them and that about 200 pounds a year are pickled and salted at Martha's Vineyard."

We find many references to swordfishing from New England ports in the early nineteenth century, but evidently the Canadians didn't catch or eat swordfish until the turn of the twentieth. In the 1903 Canada Department of Marine and Fisheries' *Annual Report*, we read:

> A new industry sprang up here this year in the catching of sword fish and quite a number were caught. The catching of these excellent fish has been an industry for a number of years on the coast of the United States, but it has never been followed here. It was discovered here this year that these fish were unusually abundant in our waters,

and as the price is usually a good one, our fishermen fitted out with harpoons and other appliances to capture them with the result that quite a number were taken and another year will probably see an important business done if the swordfish are as numerous as they were this season. They are among the best of the edible fishes, as all who have tasted can testify.

Outside of New England, though, not many people were acquainted with the firm, white flesh of the swordfish. Before she wrote *Silent Spring* and *the Sea around Us*, marine biologist Rachel Carson (1907–64) was editor-in-chief for the US Fish and Wildlife Service. In a 1943 report, she discussed some of the little-known seafoods that could be eaten during wartime food shortages:

> The swordfish ranks high among the "quality fish" of New England. Thick steaks entirely free from small bones are cut from this large fish. They are excellent when broiled, and planked swordfish is a special delicacy. The flesh is something like halibut in consistency, but it is more oily and has a rich, indescribable flavor that is different from that of any product of the sea. The vitamin content of swordfish liver oil is exceptionally high.

Even in the days before industrial fishing, commercial fish catches were measured in pounds or tons. Obviously, you cannot count the actual number of codfish, herring, or mackerel that are caught, so measurement by weight was considered a convenient requisite. Among other things, this reduces (or enlarges) the fish to a *commodity*, which makes it easy to forget that the "product" was once a living creature. In their 1953 synopsis of the fishes of the Gulf of Maine, Bigelow and Schroeder wrote, "Our only clue to the numbers of swordfish that visit our waters is the poundage landed yearly. The smallest year's catch reported landed at Portland, Gloucester, and Boston, within the period of 1904 to 1929 was 833,000 pounds (in 1919), the largest was 4,593,000 pounds (in 1929), the average about 2,000,000 pounds or anywhere between 4,000 and 18,000 fish per year. And the landings in New England ports ran from 1,715,000 to 5,070,000 pounds during the decade 1930 to 1939." Bear in mind that these catches were all accomplished one fish at a time; every one of those 18,000 fish was spotted, harpooned, brought aboard a fishing boat, and prepared for market. As Jordan and Evermann write, "Practically all the swordfish brought in to market are harpooned, we have never heard of one caught in net or seine, nor is it likely that any net now in use would hold a large one."

In the early days of the New England swordfishery, between 1883 and 1895, the average dressed weight of swordfish, according to Bigelow and Schroeder, was "between 200 pounds and 310 pounds, falling to 114–186 pounds for the years 1917, 1919, 1926, and 1929–30. . . . A 7-foot fish weighs abut 120 pounds; 10 to 11-foot fish about 250 pounds; fish of 13 to 13½ feet, about 600 to 700 pounds." (All measurements include the sword, which made up approximately one-third of the fish's total length.) There once were tens of thousands of swordfish swimming around the Gulf of Maine, and each of them required hundreds of thousands of baitfishes to sustain them—the Gulf of Maine must have been wall-to-wall fishes. (Never mind the uncountable planktonic animals that were required to sustain the baitfishes.) In an earlier ocean unpopulated by fishermen, the swordfish was at the peak of the food pyramid; it was the apex of apex predators, dominant and unthreatened. Then came the harpooners.

The quarry was spotted at the surface, chased and then harpooned, so swordfishing was not unlike whaling (although the fish were considerably smaller); even the gear was similar. In whale hunting, unlike most kinds of fishing, the quarry had to be *seen* before it could be caught. In Yankee whaling, the lookout stood high in the rigging, searching for the telltale blow of a whale, and when one was spotted, the boats were lowered. In swordfishing, the intermediate steps of lowering the boats and harpooning the quarry at sea were eliminated, because the swordfish could be harpooned directly from the "pulpit," a special platform at the end of the bowsprit, where the "striker" either threw the harpoon or stabbed the fish directly from his platform. Like that of the whaler, a swordfish harpoon consisted of a long pole fitted with a socketed iron rod that terminated in a double-arrow point. The harpoon head was warped to a line that was 40–80 fathoms in length, on the end of which was a barrel or buoy. When the fish was darted or struck, the dory was lowered, and the doryman snagged the line and hung on to the pain-maddened fish. This often resulted in the equivalent of a "Nantucket sleigh ride," where a half ton of wounded swordfish, instead of a whale,

Figure 1.1. In the 1880s, fisherman harpooned swordfish from an iron hoop mounted on the bowsprit. (Notice the good-luck horseshoe). Illustration from G. B. Goode, ed., *The Fisheries and Fishery Industries of the United States* (1887).

towed the dory until the fish tired, and then, like the exhausted whale, was finished off with a lance thrust. The whale was towed back to the ship for flensing and processing; the dead swordfish was hauled into the dory and brought back to the boat. "The sword, fins and tail are cut off immediately," wrote Tibbo, Day, and Doucet, "as each fish is hoisted on the deck of the vessel. The remainder of the dressing—cutting off the head, scrubbing the body cavity with wire brushes, washing with sea water, wiping with burlap cloths to remove loose filmy tissue, and icing—is done at the end of the day's fishing."

Because the swordfish is the ultimate catch, every fisherman wants one. Peter Matthiessen, conservationist, traveler, novelist, and above all, America's most eloquent writer on nature and ecology, recounted the time he was out on his boat *Merlin* and heard a radio report that

Figure 1.2. Famed marine author and illustrator Clifford Ashley (1881–1947) painted this dramatic picture of New England swordfishermen at work. Richard Ellis collection.

a "striker" had harpooned 30 swordfish south of Martha's Vineyard. He headed *Merlin* for the spot: "And so I was wound tight with expectation as I ran forward to the pulpit, freed the long harpoon lashed across the rails, and stared ahead at the two curved blades tracing a thin slit on the water. The beautiful fish was of moderate size, less than two hundred pounds, a swift and graceful distillation of blue-silver sea (larger fish are darker and look brown). Its round eye, a few inches below the shining surface, appeared huge. I was still staring when the night-blue fish shivered and shot away, leaving only the deep sun rays in the sea." The striker who was with Matthiessen tells him, "I put you right on that fish, but you never struck him, and you know why? You seen the eye. My old man taught me never to look at the eye, just at the dorsal fin. . . . Nobody believes how big that eye is, and by the time they get over the surprise, the bow is past him and that fish is gone."

This big fish was spared, at least for the moment: like most of the general population of swordfish, little threat was posed to it by the fishing techniques used prior to the introduction of the longline. Even the subtraction of two million pounds (1,000 tons) per year of swordfish did not appear to have a noticeable effect on the population. As recently as 1961—just before the introduction of longline fishery—Tibbo et al. wrote,

> Although there is no measure of the abundance of swordfish in the northwest Atlantic, there seems to be little reason for concern about the supply. The fish taken are nearly all large adults and only occasionally are small fish caught. There is no evidence of any substantial changes in the abundance of swordfish in recent years. The harpoon method of fishing probably takes only a small fraction of the fish in an area as each individual fish must be the object of a special pursuit. The solitary habits of the swordfish also protect it from wholesale capture as is the case with closely schooled fish.

Carl Safina, who wrote so movingly about the decline of swordfish in *Song for the Blue Ocean*, brilliantly describes swordfishing in a later book, *Voyage of the Turtle*:

> We turn, boat and swordfish, two of the ocean's most self-assured top killers, approaching each other. We're executing a strange ambush in the form of a direct approach. Surely it hears our hull, but the Swordfish has no natural predators the size of our boat. . . . At about five to six knots we're rapidly closing the gap. The bruise-purple swordfish makes no course alteration and in no way reacts. Its bill astonishes with length and breadth. . . . Franklyn remains relaxed, harpoon in hands. Suddenly and momentarily he is poised directly over the fish's back, traveling nearly past the swordfish

> when he thrusts into the spot, driving his bronze harpoon-head deep into flesh. Nothing of what I expect ensues. No explosion of water. No lashing of the rapier bill, no counterthrust or desperate backslash. No yelling. The stricken fish merely turns sideways as though stunned, its steely side and the blank gaze of its enormous eye flashing upward. Franklyn has driven the harpoon so deep into the animal's thick back it has made a wound in its belly.
>
> The point detaches from the shaft and the fish angles downward, pulling 600 feet of rope and buoys attached to the dart deep within. We will pick it up later, when it's safe to do so.

Safina was aboard the swordfishing boat *Joel Troy* on Georges Bank with Franklyn d'Entremont, "master swordfisherman of near-legendary local repute." Safina continues:

> There's virtually no harpooning anymore in U.S. waters, nor in most of the rest of Canada. Last of a breed, these guys are true masters at a dying craft. Swordfish harpooners are just about the last people on Earth who engage big game with a hand-thrust spear. They follow a long series of peoples whose hunting cultures have gone extinct, withered to oblivion as they drove their prey to depletion or the world changed around them, inter alia: mammoth hunters, buffalo hunters, coastal whalers, Inuit sealers, Maori moa hunters, Maasai lion hunters, swordfish harpooners. High priests in a vanishing act. Last men standing. Same damn thing.

Voyage of the Turtle is mostly about the plight of the various species of sea turtles, but when Safina boarded the *Joel Troy* to search for leatherbacks on the Grand Banks, he found himself on a working swordfish boat. Safina is a respectful fisherman as well as a professional conservationist, so he has no problem watching swordfish being harpooned. Still he writes, "One can only hope the swordfish is somehow incapable of feeling the torment we've inflicted as it bears its strange burden of rope and floats. The fish will tow until the the combined exhaustion, suffocation, and blood loss kills it. . . . There is no fighting the great gladiator as in the old days, when each fish took a deployed boatman on a 'sleigh ride' before he wrestled it to the surface, killing it with lance-stabs to the heart and gills. . . . These days, the gladiator dies alone in a sphere of darkness, hundreds of feet below the sparkling surface."

The Strait of Messina is the narrow passage between the eastern tip of Sicily and Calabria, the toe of the Italian boot. At its narrowest point, the strait measures less than two miles in width, but near the town of Messina, it is more than three miles wide. The maximum depth is 250 m (830 ft.). The conjunction of the Tyrrhenian and Ionian Seas makes for particularly strong currents, leading to unique fishing methods that are still in practice today. As reported by the Greek historian Polybius (ca. 200–118 BC) more than 2,200 years ago, the *pesce spada* were originally located by spotters on the hills overlooking the strait on the stretch of coast between Scilla and Palmi on the Calabrian side.[2] Each one of these spotters had his own area to control, with fixed boundaries. They indicated the presence and position of the fish to the fishermen in boats, who hunted them down with harpoons.

Later, the swordfishermen took to the sea in small, fast rowboats, supported by a big boat with a mainmast about ten meters high. When a sailor spotted a swordfish from the mainmast, he signaled to the man in the bow, who hurled a long wooden harpoon at the fish. Over time, the system remained essentially the same, but the boats and the technology were improved. It certainly takes sophisticated engineering skills to build the incredible rigs used by these Sicilian fishermen, but the plan is still "spot fish; harpoon fish; kill fish." Now inboard-powered fishing boats (*felucca*) about 40 feet long have a 50-foot vertical mast set amidships, accessible by a ladder that it part of the mast's construction, with a crow's nest at the top. The crew is usually composed of six men, three of whom are spotters high above the surface of the sea. A

2. The town of Scilla in Calabria overlooks the tempestuous Strait of Messina and is a contributing element in the myth of Scylla and Charybdis, where Scylla was a hazardous rocky shoal and Charybdis was a deadly maelstrom. These hazards were so close to each other that avoiding Charybdis meant passing too close to Scylla and vice versa. According to Homer, Odysseus was forced to choose which to confront while passing through the strait; he chose to bypass Scylla and lose only a few sailors, rather than risk the loss of his ship in the whirlpool. Having to navigate between two hazards eventually entered idiomatic use, as in the English seafaring phrase, "Between a rock and a hard place."

Figure 1.3. Two Sicilian swordfish boats in the harbor. Note the towering mainmasts and the harpooner's incredible bowsprit. Photograph courtesy of Silvestro Greco.

Figure 1.4. Poised to throw the harpoon. Photograph courtesy of Silvestro Greco.

Figure 1.5. The fish is collected in a small boat. Photograph courtesy of Silvestro Greco.

Figure 1.6. The catch aboard the swordfish boat. Photograph courtesy of Silvestro Greco.

complete set of controls for driving the boat is in the crow's nest, so the spotters (and the steersman) have an enormous scope of vision as they search for swordfish at the surface. When he spots a fish, the steersman gently heads the boat in its direction, and the harpooner silently edges out along a bowsprit-platform that is perhaps 100 feet long, considerably longer than the tower is high. The stalking of a fish that the fishermen believe is sleeping—and it may very well be—takes place as quietly as possible, until the man on the bowsprit hurls the *arpione* at the unsuspecting fish. The fish is then hauled aboard the boat, where it will be killed and prepared for market. How many swordfish are killed in this fishery? Hard to say. According to a 2005 study of this fishery by Di Natale, Celona, and Mangano, "Data from this fishery are very difficult to collect, because all fishes are sold on the local market and are often landed just after each catch, because no vessel is equipped with a refrigerator and the fishing area is often not far from the coast."

2 Before the Swordfish

Swordfishes are large oceanic fishes, occurring in all tropical and temperate seas. They are endowed with extraordinary strength and speed. Next to those powerful and masterful cetaceans the Killers, I think they may be regarded as Lords of the Sea, and they are the largest of all teleostean or bony fishes, attaining lengths of from 12 to 15 feet. The sword, spear, or bayonet, as the case may be, is a prolongation of the snout. The sword of a typical swordfish is shaped like a two-edge Scottish broadsword and may reach a length of as much as 4 feet 6 inches.—D. G. Stead, *Giants and Pigmies of the Deep*

The current crop of billfishes is not the first group of marine creatures to develop a long upper jawbone and probably won't be the last. In the evolutionary process known as convergence, where unrelated species develop similar modifications that allow them to function in more or less the same fashion, there were several marine animals that were equipped with some form of forward-projecting "sword."

If a casual observer looked at a drawing of an ichthyosaur and identified it as a dolphin, the mistake could be easily forgiven. A closer examination would reveal a pair of hind flippers and, more critical, that the tail fin, instead of being horizontal as it is in cetaceans, is vertical as it is in sharks. Those differences notwithstanding, the ichthyosaurs of the Middle Triassic through the Cretaceous demonstrate an incredible example of convergent evolution, with similar traits developing in totally unrelated groups of animals. In this case, the ichthyosaurs would adapt to a life in the ocean, flourish, and die out 40 million years before the earliest cetaceans. In shape, they were streamlined like the fastest fishes, and they breathed through nostrils located above the eyes, not on the tip of the snout. The ichthyosaurs used their lunate caudal fins for propulsion, much the way sharks do today. They were reptiles and therefore had well-developed ribs from their necks to their tails—some of the post-Triassic ichthyosaurs had more than 80 pairs of ribs—which suggested a rigid trunk, with propulsion deriving from undulations of the vertical tail fin.

The ichthyosaurs lived in the early Triassic (about 250 million years ago) and went extinct about 93 million years ago. During their 150-million-year history, they developed many varied forms, all of them conforming to the same basic body plan, with thematic variations. Some were small and snub-nosed; others had long, pincerlike jaws; while others had an overhanging upper jaw like that of a swordfish. The earliest ichthyosaurs were smallish, no more than three feet long, but there were some monsters that reached a length of 50 feet, and some were even larger. We know that they were all aquatic because, as Michael Taylor points out, "Paleontologists find ichthyosaur fossils only in coastal or marine rocks." Besides, unless it was a snake, an animal without functional legs would have had a rather difficult time on land. (Living sea turtles, also with four flippers, do not move on land very well.)

The southwestern coast of England has been intensively scoured for fossils since young Mary Anning found the first ichthyosaur in 1811. In Somerset in 1984, an ichthyosaur was found whose upper jaw extended a considerable distance beyond the lower, or as McGowan describes it, "mandible shorter than skull but exceeding 60% of skull length. Snout extends well beyond anterior tip of mandible but length of snout . . . not greatly exceeding length of mandible." McGowan named it *Excalibosaurus*, both for its swordlike jaw and for the fact that it was found in the west country, the place of the legendary appearance of King Arthur's sword. The first of the swordfish-ichthyosaurs was described by Dr. Gideon Mantell in 1851, from the Upper Liassic of Whitby, Yorkshire. He named it *Ichthyosaurus longirostris*, but it was later determined that it differed enough from the other known *Ichthyosaurus* species to warrant its own genus, and so it became *Eurhinosaurus longirostris*, the "broad-nosed lizard with a long beak." *Eurhinosaurus* had long, slender pectoral fins and a tail that was probably lunate like that of the broadbill swordfish *Xiphias gladius*. Today's swordfish has a single, fleshy anal fin, but like all ichthyosaurs, *Eurhinosaurus* had a pair of hind flippers. Other significant differences included the absence of gills in the ichthyosaurs (they were air-breathing reptiles and had to surface to breathe, whereas fishes breathe water) and the presence of teeth in the jaws of the ichthyosaur. (The swordfish is toothless.) The body plan of the two is so similar that a layperson shown silhouettes of *Eurhinosaurus* and *Xiphias* would be hard-pressed to differentiate one from the other. The ichthyosaur was considerably larger than the fish, however; swordfish can reach a length of 15 feet (including sword), but some fossils of *Eurhinosaurus longirostris* are more than twice that length.

McGowan recognized the usefulness of swordfish studies in relation to the long-snouted ichthyosaurs and therefore studied swordfish in an attempt to learn how *Eurhinosaurus* might have earned a living. Most ichthyosaurs had teeth, and they undoubtedly used them to catch their prey. It is not clear, however, what purpose was served by the teeth in that part of the upper jaw of *Eurhinosaurus*. Because of the severe overbite, the upper teeth could not make contact with the teeth in the lower jaw or, for that matter, with anything at all. If the function of an elongated rostrum (which in Latin means beak, bill, or snout) has to be conjectured in the living swordfish and marlins, imagine the problems involved in trying to figure out how an animal that has been extinct

Figure 2.1. The 100-million-year-old ichthyosaur *Eurhinosaurus* (*top*) bears remarkable resemblance to the modern swordfish (*bottom*), but unlike the swordfish, the ichthyosaur had teeth (and two pairs of flippers). Drawing by Richard Ellis.

for 180 million years might have used its elongated upper jaw (or its teeth). How extinct reptiles solved their eating problems, however, has little or nothing to do with the billfishes. Some ichthyosaurs were elongated, fast-swimming marine predators with elongated upper jaws, but teeth or no, they were air-breathing *reptiles*, while the marlins and swordfishes are *fishes* and breathe water with gills.

The swordfish and the sawfish are very large fish with a pronounced extension of the upper jaw. A simple look at the name, however, will eliminate any confusion: where the swordfish possesses a *sword*, the sawfish has a *saw*, a horizontally flattened blade studded with large, evenly spaced teeth on its outer margins. As with the mystery of the sword, it is not immediately evident how this creature might employ its weapon, but it is not used to saw its prey in half, and stories of the sawfish cutting large lumps of flesh from the bodies of fish are equally ridiculous, although the otherwise dependable Norman and Fraser noted in their 1938 *Giant Fishes, Whales and Dolphins*, "On occasion it may even attack larger fishes, cutting large lumps of flesh from their bodies with the 'saw,' but stories of sawfish attacking whales are probably without foundation." (The authors also repeat a story of an Indian "Saw-fish [that] had once cut a bather entirely in two," but this must also be treated as a baseless exaggeration, with no semblance of fact.)

Still, a 20-foot-long fish equipped with what appears to be significant offensive armament is going to engender a lot of fish stories. For example, in *Sea and Land*, his 1887 compilation of animal folktales, J. W. Buel recounts the tale of a "saw-fish" caught by a Dr. Quackenbush of Mayport, Florida. Upon feeling a tug on his line, the fisherman, who was accompanied by his nine-year-old daughter, tried to reel in the unknown fish, when, "suddenly shot up a saw-bill four feet in length, and began striking, apparently blindly, from side to side until it reached the boat, when, with one terrific stroke, it tore away two feet of the stern down to the water's edge. . . . With admirable presence of mind, [Quackenbush] drives an oar into the infuriated fish's mouth with such force as to repel the vicious attack." (When Buel wrote, "It grows to a length of twenty feet, but the powerful, keen, and heavily dentilated [having teeth] blade is one-third of its entire length," he thought he was talking about the swordfish.

The sawfish is considerably less antagonistic than Mr. Buel would have us believe. Sawfish are actually slow-swimming bottom-feeders, usually found close to shore in shallow water, where they use the saw for digging up sand and mud to uncover buried shellfish (the rostrum of the sawfish is sensitive to weak electrical impulses emitted by animals buried in the sand, like the head lobes of the hammerhead shark), although they have been known to slash at a school of mullet to injure or kill them. The sawfish (there are several species of the genus *Pristis*), in fact, is a shark-shaped ray, one of the cartilaginous fish that, for the most part, are characterized by a flattened rhomboid shape, with gills and mouth on the underside of the head. The pectoral fins are broad and emerge directly from the head, a sure indication of its affiliation with the rays. It has two large, evenly matched dorsal fins and a broad, sharklike tail fin with no caudal notch. Sawfish are live-bearers, but when the young are born, the saw is soft and rubbery, with the teeth embedded in it.[1]

Because sawfish are not considered game fish, there are few records kept of their maximum length or weight, but they have been reliably reported at 16–18 feet in length and a weight of 800 pounds. They can be found in shallow waters along both coasts of Florida and Central America, throughout the Caribbean, and as far south as Brazil. The really big ones, it seems, are found off Panama.

In the early 1920s, British sportsman F. A. Mitchell Hedges set out on a two-year fishing expedition, which he chronicled in 1923 in *Battles with Giant Fish*. Among the fish he claimed to have battled were a 40-pound snook, a 98-pound jack, a 1,300-pound hammerhead shark,

1. Sawfishes are rays that look like sharks; sawsharks (*Pristiophorus* spp.) actually *are* sharks, defined by the location of their gill slits on the side of the head. The flattened rostrum is armed with toothlike denticles, and their teeth are more pointed and sharklike than those of the sawfishes. Sawsharks reach a length of five feet, and with the exception of the Caribbean species *Pristiophorus schroederi*, they are all found in Indo-Pacific waters. They are deepwater sharks, found at depths of 200 or more feet. All sawsharks have a pair of barbels located about midway on the underside of the "saw." There is one species, *Pliotrema warreni*, that has six gill slits and is commonly known as the sixgill sawshark. (Most shark species have five gill slits on each side, three species have six, and one has seven.)

Figure 2.2. What a difference a name makes. Where the swordfish has a flattened sword, the sawfish has a tooth-studded saw. Drawing by Richard Ellis.

a 1,460-pound "shovelnose shark," a great white shark, a 200-pound "porpoise" (based on the photograph, actually a juvenile bottlenose dolphin, which he believed was a kind of fish), and finally, somewhere off Panama, a giant sawfish, harpooned by one of his "native boys." As the fish approached the shore, Mitchell Hedges and his companion, Lady Richmond Brown ("a damned good sportsman"), splashed through the surf and hauled it in with a stout line. Then Mitchell Hedges "smashed home two bullets where I thought its heart might be," killing it instantly. They measured the brute, and it was 24 feet 6 inches in length and 17 feet 6 inches in girth. It weighed, he says, 1.75 tons. But it was a bantamweight compared with the "leviathan of the deep" that Mitchell Hedges captured on his last day of fishing.

This monster towed Mitchell Hedges and company all over the sea for more than six hours, but this time they managed to fasten the line around some rocks on shore. Once again, Mitchell Hedges shot the fish, which, in its death throes, "reared up and then smashed straight down flat with a terrific spank on the water, while the brute gave two or three convulsive shudders." He tells us that "this brute was thirty-one feet long and weighed 5,700 pounds." Measuring a 31-foot-long fish on a remote beach in Panama is not that difficult, but how could the trigger-happy Mitchell Hedges possibly have weighed it? Entirely as a function of its dangerous-looking forward protrusion, the inoffensive sawfish has acquired a totally undeserved reputation for ferocity. In Mr. Mitchell Hedges's silly book, the weight and size of virtually every fish he claims to have caught is far beyond the known range for these species, but his stories of gigantic, thrashing sawfish are the wildest of all.

As with so many of the big fish, the sawfish is currently in trouble. Although there was never a directed fishery for them, the toothed saws have always made desirable souvenirs, and whenever one was caught—accidentally or on purpose—the saw was made into a trophy. A nineteenth-century survey identified sawfish as among the most abundant species in the Indian River system of Florida; one mullet fisherman reported catching 300 in a single season. So many sawfish were trapped in the nets of shrimp trawlers in the 1960s in the Gulf of Mexico that these accidental catches soon dropped to zero. "Even among biologists," writes Janet Raloff, "these fish were never on the radar screen . . . so their virtual disappearance in the 1970s went unnoticed." The same madness that has doomed so many sharks has also been responsible for the catastrophic decline in sawfish numbers around the world: The fins can be made into shark-fin soup. Because they are now rare, the fins of sawfish can command higher prices than those of sharks; a pair of the matched dorsal fins might be worth $3,000. In 2007, all international trade in sawfish was banned. It remains to be seen, however, if the treaty can protect the sawfish from those entrepreneurs who continue to catch them for their valuable fins and their toothy rostrum.

There were fish with swords—or at least swordlike protrusions—long before the modern *Xiphias gladius* arrived on the scene. Consider the long-extinct *Protosphyraena*. (*Proto* is Greek for "first"; *sphyraena* is Greek for a pike-like fish.)[2] A number of fossils were collected in

2. Today's barracuda is *Sphyraena barracuda*. When the first *Protosphyraena* fossil was named by Gideon Mantell from a fossil found in England in 1822, he believed it to be a relative of the barracuda. The teeth alone would support such an affiliation, but *Protosphyraena* is not only not a barracuda, it is not a swordfish either. It belongs to the extinct family Pachycormidae, which was characterized by the serrated pectoral fins, reduced pelvic fins, and a bony ros-

1872–73 in the Kansas chalk beds by B. F. Mudge, who sent them to Edward Drinker Cope for examination. Cope opined that *Protosphyraena* grew to a length of at least 10 feet and had a pronounced extension of the upper jaw like that of a marlin, but it was further armed with a prominent set of teeth. And rendering it even more deadly, *Protosphyraena* had long, scimitar-like pectoral fins that could be used as slashing weapons. In his 1917 discussion of fossil collecting in Alberta, Charles H. Sternberg described this creature thus:

> I used to think that the man-eating sharks off the Florida coast were the most blood-thirsty of the order, but this one is still worse. Notice the head is prolonged in front into a long bony snout, or ram. On account of this, I called it a snout-fish, when I first discovered their bones in the Kansas Chalk. The ram ends, you notice, in a sharp point eight or ten inches long. Then at the end of the mouth there are four lance-like teeth projecting forward and outward. The object of these is to cut wide the breach the ram makes in the quivering flesh of a mosasaur, so he can force his head into the bleeding flesh to the eye rims. But his most terrible weapons are his pectoral fins. See, they are four feet long and serrated on the cutting, or outer edge, enameled, and as sharp as a knife. They can be locked and stand out straight form the body. A sudden swing would, if he was close to a mosasaur, cut a gash several feet long in its vitals. See these fins span over eight feet. I pity the fish or reptile that comes his way.

In his 1874 discussion of *Protosphyraena*, Mudge wrote:

> The most remarkable species of fish which we have found, the present season, are of a genus new to me, and I think to science. They are armed with a long, strong weapon at the extremity of the upper jaw, something like that of a sword-fish, but round and pointed and composed of strong fibres. The jaws are provided with three kinds of teeth. On the outer edge is a row of large, flat, cutting teeth, somewhat resembling those of a shark. Inside, and placed irregularly, are

---

trum. Another pachycormid was the filter-feeding *Leedsichthys*, at an estimated length of a hundred feet, the largest fish that ever lived.

small, blunt teeth; while in the back portion of the palate is the third set—small, sharp and needle-like in shape, forming a pavement. The jaws are also fibrous, like the snout. There are three species of this genus. Prof. Marsh has them for critical scientific examination.

At least 14 specimens were collected by Mudge in 1874 from Ellis and Rooks counties in the Kansas chalk. In *Oceans of Kansas: A Natural History of the Western Interior Sea*, Mike Everhart discusses all the known fossils of *Protosphyraena*, including the 1988 discovery by his wife Pam of a complete skull and fins of *Protosphyraena nitida* in Ellis County.

Protosphyraena has been is extinct since the Late Cretaceous, about 65 million years ago (around the same time that the last of the terrestrial dinosaurs disappeared), but the idea of a fish armed with a spear lives on. The fossil record, of course, can only reveal a small sample of what lived where and when and, obviously, cannot show what did *not* live where and when. We cannot say with any degree of certainty that there were no rapier fishes in the Late Cretaceous—only that we haven't found any fossil evidence to indicate that there were. In fossil beds that can be dated around 60 million years ago, in the period known as the Late Eocene, paleontologists have found evidence of the first true billfishes.

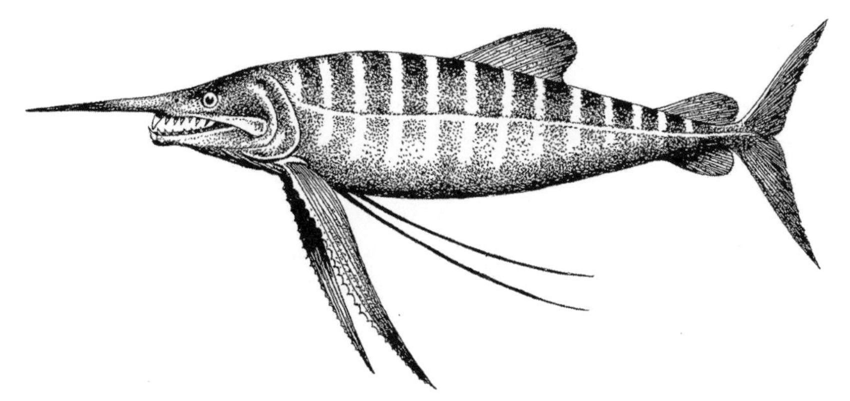

Figure 2.3. Extinct for 65 million years, Protosphyraena had the sword of a swordfish, the teeth of a barracuda, and serrations on the leading edge of its pectoral fins, which added another element to its arsenal. Drawing by Richard Ellis.

Working in Turkmenistan, Russian paleontologists E. K. Sytchevskaya and A. M. Prokofiev unearthed a 15-inch-long fragmentary fossil that they published as a new genus of early swordfish, which they named *Hemingwaya sarissa* (*Sarissa* is Greek for "spear.") Although they wrote that "the genus is named after E. Hemingway," they did not expand on this statement. (It is probably safe to assume that either Sytchevskaya or Prokofiev was a fan of Hemingway's writing about big spear-nosed fishes.) *Hemingwaya* is considered a billfish because, as Fierstine wrote to me, "it has many characters found in other early billfishes, such as, an elongated rostrum that is slightly longer than the lower jaw, both rostrum and jaws bear small villiform teeth, elongate body, the ventral (pelvic) fins are absent, and the first anal and dorsal fins are elongated. The caudal rays overlap the hypural plate (this makes the fin more stiff) and the [caudal] fin is forked." Fierstine further wrote that "it is a reasonable guess that [this] is the earliest billfish."

Now retired from a distinguished career at California Polytechnic State University at San Luis Obispo, California, Harry Fierstine concentrated on the paleontology of billfishes, which, as of a 1974 study, he divided into three families: Istiophoridae, Xiphiidae, and Xiphiorhynchidae. (For those names that begin with an *X*, use a *Z* for pronouncing it: *Xiphias* is pronounced "Zif-ee-us.") For purposes of scientific continuity, fossil species are rarely (if ever) given common names, and therefore we must be prepared to use the often difficult Greek- or Latin-derived binomials, such as *Aglyptorhynchus maxillaris* or *Xiphiorhynchus kimblalocki*. (Even *Tyrannosaurus rex*, by far the most popular of the terrestrial dinosaurs, is known only by its scientific name.) In 2006, Fierstine updated his analysis of fossil billfishes and included a detailed study of the family Paleorhynchidae ("ancient beaks"), which comprises three genera, *Homorhynchus*, *Paleorhynchus*, and *Pseudotetrapterus*. They ranged in size from about two feet to 12 feet, and all had protruding upper and lower jaws of equal length.

In his 2006 study, Fierstine raises this most interesting question: "Is there a functional explanation for extant billfishes having their rostra extend well beyond their lower jaws, whereas most extinct billfish genera have upper and lower jaws of equal length?" He answers by saying that if the rostrum is used to impale prey, it is difficult enough to remove one jaw from the impaled prey, but to remove two, "the billfish would need to withdraw its jaws from the object fairly quickly in order to pass water over its gills for respiration." In other words, the development of a single impaling weapon was an evolutionary improvement over the two-equal-jaws model because matching upper and lower jaws would have been maladaptive, and those billfishes with a more efficient method of impaling prey (and surviving the impalement) would be more likely to pass this trait to future generations. Which of course is exactly what happened: every living billfish has a longer upper jaw than a lower, most exaggerated in the living swordfish, *Xiphias gladius*. Whereas the sword in the marlins and the sailfishes might account for a quarter or less of the fish's total length, the mighty spear of the swordfish can be a third of the fish's overall length.

The Istiophoridae are represented today by the roundbills—marlins, sailfish, and spearfishes—creatures whose beaks are round in cross section and further characterized by a longer occipital region and the presence of a predentary bone at the tip of the lower jaw. Our friend *Xiphias gladius* is the sole surviving member of the Xiphiidae; and the Xiphiorhynchids, believed to be intermediate to the two extant forms, are all extinct. (*Xiphiorhynchus* can be translated as "sword-nose.") All the Xiphiorhynchid fossils had been recorded from Europe, but in 1974, with Shelton Applegate of the Los Angeles County Museum, Fierstine described the first species of a Xiphiorhynchid billfish from North America. Of the new species, which they named *Xiphiorhynchus kimblalocki* after Kim Blalock who discovered the specimens in an Eocene deposit in Clark County Mississippi, they wrote, "We speculate that *Xiphiorhynchus* is an extinct offshoot from an unknown pre-Eocene common ancestor between Xiphiidae and Istiophoridae and is closer to the Istiophoridae than the Xiphiidae."[3] In other words, in certain anatomical details—including the round cross-section of its bill—this creature resembled a marlin more than a swordfish.

3. The halfbeaks (family Hemiramphidae) are fishes in which the lower jaw is considerably longer than the upper, but these small, schooling fishes are plankton and algae-feeders not impalers. Why a protruding lower jaw would be an advantage for plankton eaters has not been explained. Ironically, the halfbeak species known as the ballyhoo (*Hemiramphus balao*) is a very popular baitfish for big-game fishermen in pursuit of marlins and sailfish.

Even though they had elongated upper and lower jaws, fishes of the genus *Blochius* are sometimes considered billfishes—and sometimes not. Because their rostrum was weakly constructed, write Fierstine and Monsch, "it is difficult to believe that it was used to stun or impale prey unless the prey was very small and soft." Instead, the feeding techniques probably resembled those of the needlefish, that is, taking the fish sideways in its pincerlike jaws, then manipulating it until it could be swallowed headfirst. "We doubt *Blochius* actually sucked prey into its mouth," they wrote, "because it lacks protrusible jaws and we believe the elongate dorsal fin eliminated *Blochius* as an obligate surface feeder." To date, 63 specimens have been found, all in the Middle Eocene of Monte Bolca, Italy, ranging in length from two to six feet. They were variously classified as gars, blennies, and even notacanthiformes (spiny eels), but when Fierstine and Monsch examined 60 of them, they concluded that they should be placed in a single family, the Blochiidae. "An interesting feature of the young [sword]fish," wrote Norman and Fraser in 1938, "is its resemblance to a fossil sword-fish (*Blochius*) found in the Eocene formation of Monte Bolca in Italy."

The most numerous specimens—61 of the total of 63—had been classified as *Blochius longirostris* ("long snout"), which is designated the type species because it was first described by G. S. Volta in 1796. Other Blochiids are *B. macropterus* ("small wing") and a specimen described as *B. mooreheadi*, which may not be valid. In 2001, Fierstine published the description of a new Blochiid, which he called *Aglyptorhynchus maxillaris*; it would have been 13 inches long if the tip of the bill had not been broken off. Like *Blochius*, *Aglyptorhynchus* was a slim, supple fish. Of the family, Fierstine and Monsch wrote, "We believe *B. longirostris* had a sinuous swimming pattern and engulfed its prey without manipulation by opening its mouth and swimming over its food. We infer that it was restricted to shallow water (less than 100 m) and inhabited and fed throughout the water column."

At the 2002 meeting of the Society of Vertebrate Paleontology, Fierstine announced the discovery of a new Xiphiorhynchid from the Early Oligocene (about 30 million years ago) of the Austrian Alps, the most complete specimen of *Xiphiorhynchus* found to date. It had all the characteristics that defined the genus (elongate rounded rostrum, denticles on the ventral surface of the bill, etc.), but it had a lower jaw as long as the upper, "a feature previously unknown in *Xiphiorhynchus*." As we've seen, the spearlike projection of the upper jaw of the marlins and swordfish is used for slashing and clubbing (and less frequently, for stabbing), but how did a fish feed with *both* of its jaws elongated? Was this massive creature a "tweezer feeder"? Or did it seal its jaws and use them together as a doubly effective club? When I asked Fierstine (pers. comm., November 2004), how a fish with two swords might feed, he answered, "*Xiphiorhynchus* . . . probably rarely impaled anything because the lower jaw was always in the way. On the other hand, I would think having both jaws open during feeding would greatly increase drag, however this added drag could probably be overcome by ramming water through the gills and out the opercule [gill cover]." Two years later, at the 2004 Society of Vertebrate Paleontology meeting, he described another Xiphiorhynchid, this one from South Carolina, which, from the size of the vertebrae, would have been as large as any known marlin or swordfish. In an interview published in *Science News*, Fierstine estimated that *Xiphiorhynchus rotundus* would have been "at least 5.1 meters [16.7 feet] long . . . the longest billfish in the fossil record."

In a 1990 update of his 1974 paleontological review of the three billfish families (Istiophoridae, Xiphiidae, Xiphiorhynchidae), Fierstine reiterated his separation of the three families, and again pointed out the scarcity of billfish fossils. For example, "the only fossil record of a sailfish is based in a single, fragmentary vertebra from the Upper Pliocene San Diego Formation"; there are no fossil specimens of *Tetrapturus* (spearfish) at all; and all the known marlin fossils belong to the genus *Makaira*. Of the swordfish, he said, "Fossil swordfish are rare, with most specimens coming from the Italian Pliocene . . . Other fossil records of *Xiphias* are erroneous or questionable . . . Disregarding identifications based on isolated vertebrae, the Xiphiidae is monotypic (*Xiphias gladius*) and has no fossil record before Pliocene times."

The marlins and spearfishes are equipped with bayonets, but only *Xiphias* has a great flattened broadsword protruding from its snout. Evolution does not provide creatures with unusual equipment so that they can make use of it but, rather, modifies creatures with features that they can eventually use to their advantage. Birds did not acquire wings so they could fly, for example, but rather learned to take advantage of some anatomical modifications that eventually led to flight. (Whether they first flew from the branches down or from the ground up is a subject of heated paleo-ornithological controversy.) But how did the swordfish's sword develop into the prodigious weapon we see today? Did the longer-, flatter-snouted swordfishes have an advantage over shorter-snouted ones and thus prevail through time? In their 1963 *History of Fishes*, British Museum ichthyologists J. R. Norman and P. H. Greenwood noted:

> The Common Sword-fish (*Xiphias*) is widely distributed in all warm seas, and grows to a length of fifteen and twenty feet. Its food seems to consist largely of fishes, and it is said to split large forms like the Bonito and Albacore with the sword, or to strike with lateral movements among a shoal of small fishes, afterwards devouring the stunned and wounded victims. It has, however, been suggested that the sword did not evolve as a weapon, but merely represents and extreme case of streamlining, the pointed rostrum acting as an efficient cutwater.

Did the sword develop so that the swordfish could chase and dispatch its prey, or did it chase and dispatch its prey because it had the equipment to do so?

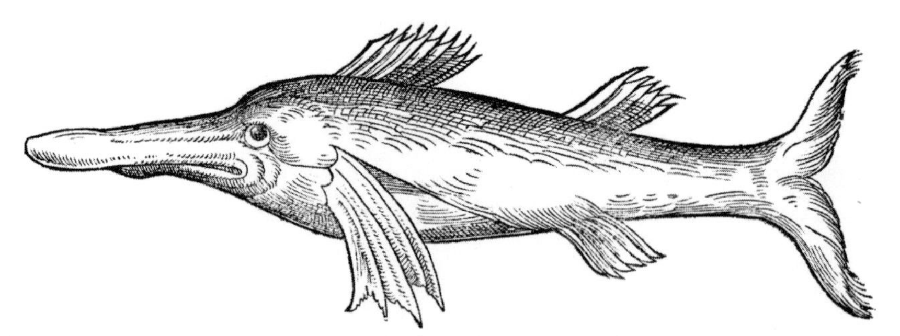

Figure 2.4. Around 1555, German naturalist Konrad Gesner illustrated the *schwerdtfisch* for his *Historia animalium*, the most comprehensive natural history written up to that time. The number and placement of the fins suggests that Gesner used a more or less accurate description, but the paddle-like snout is all wrong. Courtesy of the American Museum of Natural History Library.

3 Swordfish Biology

Not surprisingly, the swordfish's scientific name *Xiphias gladius* is derived from the sword, its most prominent feature. In Greek, *Xiphias* means swordfish, and in Latin, *gladius* means sword (it is also the root of "gladiator"), so the redundant translation of *Xiphias gladius* is "swordfish with a sword." Regardless of the native language of the speaker (or writer), the scientific name is always written as you see it here: *Xiphias gladius*. Common names, then again, may be useful in differentiating various species in various countries, but they differ from language to language. In Chinese the swordfish is called *Chien-chi-yu* and *tinmankhu*; in Spanish, it is *pez espada*, *emperador*, and *espadarte*; in Portuguese, *peixe espada* and *agulhha*; in French, *espadon*; German, *schwertfisch*; Greek, *xiphias*; Italian, *pesce spada*; Norwegian, *svedfisk*; Russian, *mechenos*; Tunisia, *bou sif*; and in Japanese, *anadaachi*, *dakuda*, *ginsazu*, *goto*, *hirakucha*, *isazu*, *kudamaki*, *meka*, *mekajikii*, *okizaara*, *rakuda*, *suzu*, *teppo*, and *tsun*. In Hawaii, it is sometimes called *shutome*, another Japanese term.

According to the *Oxford English Dictionary*, the word "fish" was originally applied to any animal living exclusively in water, such as crayfish, cuttlefish, jellyfish, or shellfish. Now we know that crayfish, cuttlefish, and jellyfish are not actually *fish*, and indeed, the term is commonly used for "any of a large and varied group of cold-blooded aquatic vertebrates possessing gills and fins." Most of the world's fishes have a bony skeleton and are classified as osteichthyes (a literal rendering of "bony fishes"); these fishes are also known as teleosts. Then

Figure 3.1. The gladiator. Photograph © Guy Harvey, Inc., 1998.

there are the sharks, with no bones at all, but rather a skeleton composed exclusively of cartilage. The *Oxford English Dictionary* also tells us that a skeleton is "the bones or bony framework of an animal body considered as a whole." "Cartilaginous skeleton" may therefore be a bit of an oxymoron, but we can get around this contradiction by employing a secondary dictionary definition: "the harder [supporting or covering] constituent part of an animal organism." (The hard [supporting] parts—the *exoskeleton*—of insects, spiders, gastropods, bivalves, crustaceans, and snails are on the outside, and of course there are no bones there, either.) The cartilaginous fishes (also known as elasmobranchs, which means strap gills and refers to their multiple gill slits) are the sharks, rays, skates, sawfish, and chimaeras. In these creatures, everything but the teeth is made of cartilage.

The largest sharks are the whale shark (*Rhincodon typus*), which can reach a length of 40 feet and weigh ten tons, and the basking shark (*Cetorhinus maximus*), which is a little smaller. Both of these behemoths are plankton-eaters, and because they would not take a baited hook, they are not considered game fishes, and in fact, they have rarely been weighed. The largest fish ever taken on rod and reel was a 2,664-pound great white shark caught by Alf Dean in South Australian waters in 1959. The next year, Mr. Dean caught another white shark that weighed 2,344 pounds. There is no question that all these fishes get larger than the records would indicate, but the International Game Fish Association (IGFA) has rules for how the fish must be caught to qualify—only one man may handle the rod during the fishing, for example—and how it must be weighed and the weight recorded. The largest billfish ever caught was a blue marlin that weighed 1,805 pounds, but several people handled the rod, so this monster was disqualified from the IGFA record book. In *Fishing the Pacific*, Kip Farrington wrote that "the largest fish I have ever personally seen was a broadbill swordfish harpooned off Chile that weighed 1,565 pounds." In the absence of IGFA restrictions, the record catches would be larger, but the books would be filled with accounts of fish caught on steel cables, or hauled in by a half a dozen people who worked together or took turns on the rod, or even—*mirabile dictu*—exaggerated the actual size of the fish.

At 40 feet in length and weighing as much as 15,000 pounds, the whale shark is by far the largest "fish" in the sea, but it is not exactly a fish. *Rhincodon typus* is a shark (albeit virtually toothless), with a skeleton composed completely of cartilage. Swordfish are not the largest of all bony fishes—if we're going by length, the flimsy oarfish, at 26 feet, can be nearly twice as long—but if weight is the criterion, the broadbill is right up there with the big boys (actually big *girls*, because all very large billfishes are females). The International Game Fish Association maintains a record book for all species of game fishes, and their book tells us that the heaviest bony fish ever caught—at least in compliance with IGFA rules—is 1,560-pound black marlin that was reeled in off Cabo Blanco, Peru, in 1953.[1] Other contenders are a blue-

1. By far the heaviest fish is the ocean sunfish or mola, a huge slab of blubber and flesh that derives its scientific name (*Mola mola*) from the Latin for "millstone." According to Gerald Woods's 1982 *Guinness Book of Animal Facts and Feats*, the largest bony fish on record was an ocean sunfish that weighed 4,928 pounds—two and a half tons. This huge relative of triggerfish and puffers feeds mostly on jellyfishes and is often seen floating horizontally at the surface,

Figure 3.2. The largest "fish" ever caught on rod and reel was this 2,664-pound great white shark, hauled in by Australian angler Alf Dean off Ceduna, South Australia, in 1959. International Game Fish Association.

fin tuna caught off Nova Scotia in 1979 that weighed 1,496 pounds and a 1,376-pound blue marlin that was caught off Hawaii in 1982. Compared to these three-quarter-ton monsters, the world's record swordfish was a relative lightweight at 1,182 pounds, but still, a fish that weighs

like a giant pancake with fins. The mola is probably the furthest thing from a game fish that it is possible to imagine.

more half a ton is a very big fish indeed. The first 1,000-pound fish—colloquially known as a grander—was a marlin caught by Zane Grey in Tahiti in 1930, and documented in *Tales of Tahitian Waters*. The fish was weighed at 1,040 pounds, even after some 200 pounds (by Grey's estimate) was gnawed off by sharks before the fish was boated. Grey estimated that the fish would have gone 1,250, but, he wrote—with uncharacteristic reserve—"best to have the record stand at the actual weight, without allowance for what he had lost."

If the term "intelligent design" had not been applied to a form of creationism that denies natural selection and attributes modifications and adaptations in living things to a thinking, planning agent (for which read: God), we might accurately apply the term to the "design" of *Xiphias gladius*. The component parts of the swordfish are integrated into one of the most efficient swimming, hunting, eating, and attacking machines ever to swim in the sea.

An adult swordfish is a graceful, tapered teardrop of a fish, with a formidable projectile point at one end and a huge, crescent-shaped tail at the other. The tail, technically known as the caudal fin, is set perpendicular to a pair of fleshy keels on the tail stock (the caudal peduncle), which are believed to impart speed and power to the tail, the fish's engine. The lunate shape and caudal keels are also characteristic of tuna, sailfish, white, mako, and porbeagle sharks, among the other contenders for the title of fastest fish in the sea. The swordfish's dorsal fin is a high sickle blade, situated far forward, and unlike the tuna, marlins, and sailfish, *Xiphias* cannot lay down this fin—it remains permanently erected, announcing its presence like a heraldic gonfalon. Its pectoral fins are long and gracefully curved, but there are no ventral fins—no plumelike appendages like those of marlins and sailfishes. The remaining complement of fins consists of a fleshy anal fin, a taglike, second dorsal, and another little fin on the underside of the caudal peduncle. These both seem to be afterthoughts—but as there are no extraneous elements in fish design, they are probably hydrodynamically important in ways we simply don't understand.

Female swordfish grow faster and live longer than males. They reach their maximum size (1,000 pounds or more) at about 15 years of age, but they are sexually mature at about the age of five. A female swordfish releases 3–16 million eggs per spawning; after the eggs are fertilized

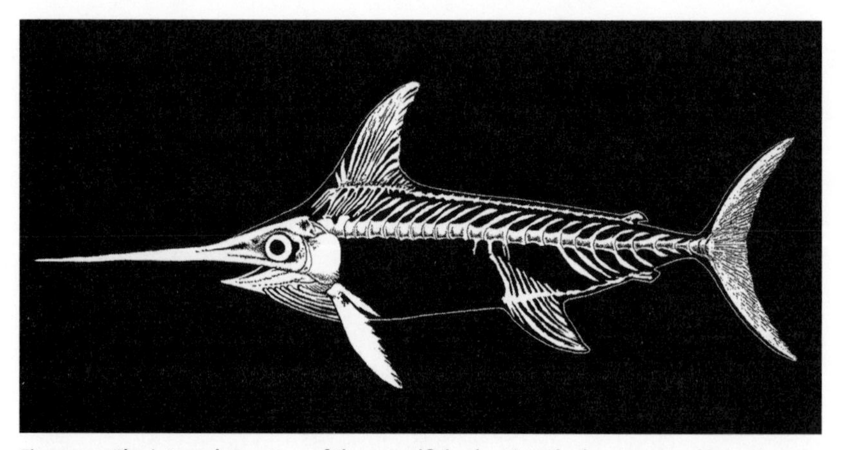

Figure 3.3. The internal structure of the swordfish, showing the huge eye socket, extra-long sword, towering dorsal fin, and powerful, lunate tail. Drawing by Richard Ellis (with help from Harry Fierstine).

they will take 2.5 days to hatch and become larvae that immediately feed and fend for themselves. On hatching, the larval swordfish is only 4 mm (1/4 inch) in length and feeds on its fellow planktons. (Anything larger than the tiny swordfish might eat it, too.) Larval swordfish are spiky little devils, called "barb-wire fortresses" by Myron Gordon, with the barbs "apparently playing a defensive role in their struggle for existence." At first, North Atlantic swordfish eat mostly tiny copepods of the genus *Corycaeus* (common in the Caribbean, Florida Current, and the continental shelf), but also planktonic worms, invertebrate eggs, and larval fishes. Swordfish larvae grow rapidly, faster than most other fishes, "and retain their larval characteristics until they are at least 188mm long [7.5 inches], a size at which most fish are considered juveniles" (Govoni, Laban, and Hare). As the baby swordfish grows, it goes through a series of morphological changes, and what begins as a full-length dorsal fin grows shorter until it becomes the tall, curved scimitar of the adult. The jaws enlarge, and then the lower jaw diminishes, resulting in a long upper jaw (the sword) and the "normal" lower jaw. Swordfish are born with teeth, proportional in size to the size of the owner, but as the fish approach maturity, the teeth disappear and the upper jaw begins to develop into a blade.

Frank Bullen (1857–1915) wrote books about sea creatures, including *The Cruise of the Cachalot: 'Round the World after Sperm Whales* (1898) and *Deep-Sea Plunderings* (1902), which was a collection of short stories about various fishes, whales, and cephalopods. In 1904 he wrote *Denizens of the Deep*, in which he wrote about "a series of lives . . . based very largely on personal observation, buttressed by scientific facts, and decorated by imagination." In *Denizens*, he seems to have been guided mostly by his decorative imagination, for his description of the swordfish sails off on wild flights of fantasy that depict *Xiphias* as "the veritable ruler of all deep-sea fish and the ruthless slayer of even the sea-shouldering whale." To Bullen, the swordfish functions as a sort of oceanic knight errant ("No ancient warrior was ever uplifted with haughtier pride in his battle plumes than *Xiphias*"), who, armed with "the terrible weapon that he carries before him," does not hesitate to use it.

Bullen envisions a domestic scene where a family of swordfishes—ten females and two males—congregate over a sandy bottom and the females lay their eggs while the males stand guard. Somehow, the eggs are fertilized, although Bullen never touches on that delicate subject, and with a flick of their tails the fish swim off after their consorts, leaving the eggs to hatch on the bottom. For the most part, Bullen's account of the life and times of the swordfish is so crazy that it's hard to know where to begin. Indeed, given that Bullen admitted to using his imagination to fill in the blank spots, it would probably be kinder to approach his work as a sort of "deep-sea fantasy," but at least in 1904, many readers believed what he was telling them. His obituary, posted by the Royal Geographic Society in 1915, called him "a well-known lecturer, and writer of stirring sea stories . . . [who] led a roving and adventurous life from quite an early age, and many of the most thrilling episodes in his books were records of his own experiences. In his latter years he was known as a successful lecturer and a writer of miscellaneous stories and articles in addition to his books."

Describing the "giant mackerel" known as *Xiphias*, Bullen immediately confuses it with the marlin and writes, incorrectly, that it "should more properly be called the Lance-fish, since the bone of the upper jaw is elongated into a weapon that is rounded and tapering to a point, more like a lance than is the flat weapon known as a sword." No longer relegated to the nooks and crannies of the reefs to feed on crabs, the maturing yard-long swordfish spots a "little band of porpoises, perfectly harmless to fish like himself, but the native blood-thirst is awakened, the lithe tail sweeps scythe-like from side to side, and in a moment he has launched himself at the flat blank flank of the sea pig, in which his sword buries itself up to his eyes nearly." Of course this is ridiculous nonsense, as is the following description of the swordfish as an adult in the next paragraph: "At last we behold him, eighteen feet long, with a weight of four hundred pounds and a sword nearly two feet in length." An 18-foot-long swordfish—if there ever was such a thing—would be a 1,500-pound *female*, and its sword would be close to six feet long.

In the world we know as reality, *Xiphias* doesn't need Bullen's romantic hyperbole; its lifestyle is more fascinating than Bullen ever imagined. In a study of the breeding habits of North Atlantic swordfish, Freddy Arocha of the Instituto Oceanográfico de Venezuela found that the females—sexually mature at a minimum length of about five feet (150 cm)—spawn south of the Sargasso Sea, in the upper Caribbean, and in the Straits of Florida (between the Florida Keys and Cuba), where the water can be 16,000 feet deep. (At 27,500 feet, the Puerto Rico Trench, just north of the island, is the deepest part of the Atlantic Ocean.) Although neither Arocha nor anyone else has ever observed swordfish in the act of spawning, it is assumed that the females lay their eggs in the company of one or more males who follow the ripe females and fertilize the eggs externally by releasing milt (sperm) into the water around them.[2] Because swordfish can carry millions of eggs at a time, they do not produce or release them all at once, but rather time

2. While fishing off Cairns, eastern Australia, Guy Harvey probably observed a spawning aggregation of black marlin: "A hundred yards away to our right, seven male marlin escorted a female of immense proportions. The group . . . were swimming fast down-sea into the current, disappearing and reappearing, flashing their neon-blue bodies on and off. The males, all approximately 200 pounds, were jockeying for position, changing colors depending on how the evening sun lit up their iridescent bodies—metallic bronzes, greens and brilliant blues and purples . . . We were able to track the group of marlin for another five minutes and we watched the males swim up alongside the big female. It was quite likely she was spawning eggs in a constant stream, as the males took up position beside her to shed their milt before dropping back to allow another male to move in."

the release throughout the breeding season, which may last for several months. Swordfish are mesopelagic fish, and most likely lay eggs in the water column with the water temperature around 75°F (24°C).

Scientists can determine the age of bony fishes by examining the otoliths—that is, accretions of calcium carbonate that develop in the ear canals. Otoliths provide the fish with a sense of balance and orientation and also aid in hearing. They acquire yearly growth rings, or annuli, much like trees do: concentric rings around Year One are at the center. The age of the fish at the time of collection can be determined by counting the annuli. Because you cannot ask a fish to submit to a CAT-scan in order to count the growth rings, you have to kill the fish to get the otoliths. Using another method of age estimation (sections through anal-fin spines), Ehrhardt, Robbins, and Arocha found that the average length for a ten-year-old female swordfish from the Northwest Atlantic was 7.5 feet (226 cm), while a male of the same age was 6 feet (178 cm). It is almost impossible to obtain a maximum life span for a particular species; the best that can be hoped for is to use the age at which a large adult fish was killed. The "oldest" swordfish ever examined was ten years old, but of course there may be older ones out there that are too sagacious to be caught.

Like tuna, swordfish have no scales, the better to pass through the water smoothly. Some regard the swordfish as the most powerful fish in the ocean; its ability to drive its sword through a foot or more a ship's oak planking supports such a contention. The swordfish's dorsal coloration has been described as iridescent purple, bronze, brownish-black, blue-gray, or dark silver, but the belly is always silvery-white. Aside from the spear, the most prominent features of the swordfish are its enormous eyes, as big as an orange, and a brilliant blue in life. Big eyes suggest a habitus of deep (or at least dark) water, and we know that *Xiphias gladius* spends some of its time at considerable depths. The celebrated encounter with the submersible *Alvin* occurred at 2,000 feet, and an eyewitness peering through the porthole reported that he "noticed a hummocky feature on the bottom . . . [which] identified itself as a large swordfish," which then charged the submersible. Was this great ocean ranger buried on the bottom? If so, why? The attack on *Alvin* occurred at 610 meters, and in his 1967 discussion of this incident, Woods Hole oceanographer E. F. K. Zarudzki refers to captures of swordfish in halibut nets near the bottom at 400 meters and also mentions R. L. Haedrich's observations of swordfish at 400–630 meters. A photograph in Zarudzki's article shows the stomach contents of a swordfish, which includes stomiatids and myctophids, deep-water species that indicate that this particular fish was feeding at 300 meters or deeper.

In 1922, long before it was suspected that swordfish spent time near (or on) the bottom, Harvard zoologist J. S. Kingsley examined several deepwater lantern fishes (Scopelidae) taken from the stomach of a swordfish that had been caught in waters more than 3,000 feet deep: "The specimens were quite fresh and the digestive juices had hardly affected the integument, and the phosphorescent organs along the sides were in good shape," he wrote and concluded that "from this it would appear that the swordfish do descend to considerable depth for their meals, and from the good condition of the specimens, it would appear that they make a rapid transition from the feeding grounds to the surface for the after-dinner siesta." Forty years later, in a more detailed study, Scott and Tibbo found that lantern fishes made up only a part of the swordfish's diet. In the western North Atlantic, they examined the stomach contents of 514 swordfish and found that mackerel were by far the most popular prey item, with lantern fishes a distant tenth.

Some of the prey species—such as mackerel and herring—are pelagic schooling fishes that live near the surface, and swordfishes spend a lot of time there, too. According to Jordan and Evermann, "They have often been described as rising through schools of mackerel, menhaden, and other fishes, striking right and left with their swords, then turning to gobble the dead or mangled fish. And we have seen them so employed on more than one occasion, to judge from the commotion." Bullen, who never saw a swordfish feed—and may never have seen a swordfish at all—describes the feeding frenzy thus: "Presently they fell in with a great school of bonito, who, taken by surprise, became utterly demoralized with fear, and huddling together in one compact mass permitted the ravenous monsters to dash again and again into their midst, with such velocity as to split sometimes four bodies into fragments at one blow."

The larger predatory sharks are usually considered the most dangerous fish in the sea because of their numerous teeth and their occasional predisposition to bite people with them. But the billfishes—swordfish, marlins, sailfishes—are the only fishes that are equipped with what truly qualifies as a *weapon.* In the marlins and sailfish—technically the Istiophoridae—it is relatively short, sharply pointed, and round in cross-section, but in *Xiphias gladius,* it is long, edged, and flattened like a broadsword. (Another of its common names is broadbill swordfish.) Indeed, where the sword of a large marlin may account for perhaps a quarter of the fish's length, the fully developed sword of *Xiphias* can be one-third of a fish's total length and often appears disproportionately long. This weapon is responsible for the wild tales about the ferocity of the swordfish—why have equipment like that if you're not going to use it?—and the early literature is filled with stories about swordfishes attacking everything from boats to whales. In *Sea and Land,* an 1887 compilation subtitled *The Wonderful and Curious Things of Nature Existing Before and Since the Deluge,*" J. W. Buel wrote:

> Next to the shark in ferocity and voraciousness is the Swordfish, a habitant of nearly all the seas, but most populous in the Mediterranean. It grows to a length of twenty feet, but the powerful, keen, and heavily dentilated blade is one-third of its entire length. Like the shark, nature has equipped the swordfish with so dreadful a weapon, that its province seems to be war, though strange enough, it is comparatively innocent and extremely timid before man, confining its ravages to its fellow-denizens of the deep.

Even today, we are not sure how a swordfish actually uses its bill. It is horizontally flattened and sharp on the edges, so it has been assumed that the swordfish enters a school of fishes and slashes wildly, cutting, stunning, or otherwise incapacitating the prey items, which it then eats at its leisure, but since few people have ever actually witnessed this activity, the use of the sword must remain conjectural. In their 1968 study "Food and Feeding Habits of the Swordfish," Scott and Tibbo wrote, "The swordfish differs from the spearfishes (marlins and sailfishes) in that the sword is long and it is dorso-ventrally compressed (hence the name broadbill) whereas the spearfishes have a shorter spear and it is slightly compressed laterally. Thus, the swordfish appears to be more highly specialized for lateral slashing. Such a specialization would seem to be pointless unless directed to a vertically oriented prey, or

Figure 3.4. *Ferocious Attack of the Sword-Fish*, by J. W. Buel, in *Sea and Land: An Illustrated History of the Wonderful and Curious Things of Nature Existing Before and Since the Deluge* (1887).

unless the swordfish slashes while vertically oriented, as when ascending or descending." But Ralph Bandini, a renowned fisherman and one of the founding members of the Tuna Club of Catalina, saw a broadbill "cut a barracuda in half, *in the water*, as cleanly as it could be done with a butcher's cleaver on the block." Bandini does not mention the angle of attack, but it is clear that the broadbill's sharp-edged sword can be used with terrible effectiveness on smaller fishes. Or on other swordfish. In the introduction to a 1981 paper on daily patterns in the activities of swordfish, Frank Carey and Bruce Robison include a fascinating note about an intraspecies attack: "We have seen penetrating wounds in swordfish and Edlin [pers. comm.] found a 15cm fragment of a swordfish bill that entered near the heart of a 70 kg [150 pound] swordfish and was driven back into the body cavity, which may indicate that they strike each other."

In contrast to almost every other suggestion about swordfish feeding techniques, in his book *Billfish*, Charles O. Mather wrote, "Essentially a bottom feeder, a broadbill is believed to use his bill as a tool to obtain crustaceans from their cracks or attachments and to enjoy crabs and crayfish," an utterly preposterous suggestion. The swordfish is not a bottom feeder; it does not use its bill as a pry bar; and it does not eat—let alone *enjoy*—crabs and crayfish. In *Blue Planet*, the book that accompanied the 2001 BBC-TV series of the same name, written by Andrew Byatt, Alastair Fothergill, and Martha Holmes, there is a lovely shot of a free-swimming swordfish, alongside which we find this caption: "The swordfish is an extremely powerful deep-water predator that probably uses its bill to dig out prey from the sea floor." Nonsense! The swordfish is one of the largest, strongest, and fastest fish in the sea. With powerful muscles that are warmed by countercurrent heat exchangers; eyes designed to pick out fast moving prey items; and a slashing, sharp-edged broadsword—the lordly gladiator is no grubber in seafloor mud.

In 1840, in his *Narrative of a Whaling Voyage around the Globe*, Frederick Debell Bennett wrote that "the sword-fish . . . subsists by making rapid darts amongst a shoal of small fish, and after transfixing as many as possible on the beak or sword . . . shakes them off by a retrograde movement or by moving the sword violently from side to side and devouring them. . . . I have seen a sword-fish thus strike and devour three bonita in a very dexterous and rapid manner." This description of a fish kabob is most unlikely, but in *Living Fishes of the World*, ichthyologist Earl Herald repeats it, writing that "the sword may be used to impale fishes during feeding." The sword of *Xiphias* is not as sharply pointed as that of a marlin or spearfish; and even if such a process could be made to work, the swordfish would be unable to get at the dead fishes stuck on the end of its nose unless it shook them off (as Bennett described), a somewhat impractical activity. (Marlins, as we shall see, do occasionally impale their victims, and then shake them off to devour them.)

By this time, there have been many studies of the stomach contents of *Xiphias*, conducted with the idea of learning what (and how) swordfish eat. The answer to "what" is easy: throughout their enormous range, swordfish eat mostly squid and fishes, in various proportions, at various depths, and at various times of the day and night. This is known as opportunistic feeding—that is, taking advantage of what is available, or, as Toll and Hess put it, "prey composition is independent of season, fish size, or sex . . . stomach contents appear to reflect the diversity and relative abundance of potential prey." Those species that aggregate are favored, but a hungry swordfish can also dispatch large fishes and cephalopods that swim alone. Even species known to be agressive predators themselves—such as the cutlassfish, the Humboldt squid (*Dosidicus*), and (maybe) even the giant squid (*Architeuthis*)—can fall victim to the well-armed swordfish.

From the plentiful evidence of punctured boats, submarines, whales, and people, there is no doubt that the swordfish can—and does—use its sword as a lance. Indeed, a powerful fish with a piercing weapon projecting from its head seems designed for just such activities. Like all other fishes, the swordfish has no "neck," so in order to move its head from side to side in a slashing motion, it would have to move its entire body. Water resistance to the sword would be minimal, but the resistance to the dense musculature of the fish's body would be substantial. Because the swordfish's body flexes with each oscillation of the lunate tail fin, the fish's head—and therefore the point of its bill—would move through a small arc as the body moves. As Harry Fierstine wrote in answer to my questions about the possible head movements of swordfish,

> Just because most fish don't have a neck, don't undersell their ability to have head movement. A swordfish has a very long bill and head, and if it could swing its head just a few degrees to the side, the tip of the bill would move quite a distance. Couple head movement with body movement and you could produce quite a force with the bill. . . . I've never manipulated the head in a fresh swordfish, but I'll bet the joint allows some lateral movement of the head. I'm currently studying an extinct billfish that has a ball and socket joint between the skull and first vertebra, and I postulate the joint allows lateral movement.

Because it is the tail oscillations that propel the fish, it would have to strike the prey items while lurching forward—which might well provide part of the answer to the question of how the various billfishes feed.

There are films where one can see sharks, dolphins, sailfish, and marlins—sometimes even working together—feeding on "balled" bait-fish. The predators herd a group of fish into a very tight ball, and the fish start to spin rapidly, like an underwater whirlwind, each fish trying to get to the center of the ball so as not to be the one that is eaten. The writhing ball makes it difficult for the predator to select an individual prey item, but having all the food bunched together gives a great advantage to the feeder. If the predator is armed with a rapier, charging the ball enables the billfish to kill many of the baitfish, and if it tosses its head while its spear is in the "ball," even more baitfish would be bashed and slashed. No one has ever observed swordfish capturing their prey, but if they attacked bait-balls, some of the mysteries of swordfish feeding might be solved. Oblivious to our difficulties in understanding (or observing) the process, swordfish obviously manage to feed themselves quite efficiently.

For the most part, evidence for the swordfish's use of the sword is circular and circumstantial: the fish has a sharp-edged sword; stomach contents often show fish pierced or chopped in half, ergo, the sword-fish must have used its sword. By this reasoning, however, most of the fish in the swordfish's stomach should be sliced, diced, or pierced, but in fact, only a small proportion show signs of physical injury. Describing swordfish caught off Nova Scotia by the Lerner Expedition in 1936, William Gregory and G. Miles Conrad wrote, "The expedition observed nothing contradictory to the common belief that the swordfish pursues the herring, mackerel and cuttle-fishes and with sharp swings to the right and left strikes them with its sword and either cuts or stuns them. It certainly often swallows them whole . . . but some were found in the swordfish stomach with broken backs." A hunting swordfish probably slashes at a school of baitfish before gobbling them up individually, but probably misses more than it hits.

Maybe the sword is used as a kind of aiming apparatus, where the fish zeroes in on prey item just before swallowing it. The videos of feeding marlins and sailfish show them charging into a balled-up school of baitfish and snapping up the prey one at a time. These sharp-snouted billfish may be wasting a good weapon, as there are plenty of other animals that can be seen feeding on the same bait-balls, and neither sea lions nor sharks use a sword. Indeed, most predatory fishes are swordless—think of that snub-nosed apex predator, the bluefin tuna—so a spike on the end of the nose might not really be necessary. All that seems to be required is speed, maneuverability, and good eyesight. Snapping up small, elusive prey items is not all that unusual; predatory fish do it and so do most bats and some birds. Bats do it in the dark, aided by the ability to echolocate their prey, but birds like swifts, swallows, and flycatchers hunt fast-moving, flying insects by sight. Bats have teeth, of course, but birds don't, and neither do swordfish; engulfing a small fish probably doesn't require teeth. In the depths, where the only source of light is the bioluminescence of deep-sea fishes and squid, slashing at the little points of light would not seem to be a particularly effective hunting technique, but spotting the little fishes by their lights and then gobbling them up probably would work and, incidentally, would go a long way toward explaining the swordfish's incredible, oversized, and overheated blue eyes.

There is another large marine animal with an even more prominent—and even more controversial—protrusion from the front of its face. The small whale known as the narwhal (*Monodon monoceros*) has a spiraled ivory tusk protruding through its upper lip. Found only in males, the tusk is actually the greatly elongated upper left canine tooth and may reach a length of eight feet. (An adult narwhal is about 16 feet long—not including the tusk.) It has been suggested that the narwhal might use its tusk to punch holes in the ice; that it is used in jousts or duels with other males; that it is used to spear fish; or that it is a secondary sexual characteristic that serves to identify the dominant males.[3] Poking holes in the ice with an elongated tooth would be rather

3. In Jules Verne's 1870 *Twenty Thousand Leagues under the Sea,* Professor Arronax identifies the possible perpetrator of attacks on merchant ships at sea as a giant narwhal, noting that "the common narwhal or sea unicorn often attains a length of 60 feet. Multiply this dimension by five, by ten even, endow this cetacean with strength proportional to its size, enlarge its offensive weapon, and you will obtain the required animal. . . . The narwhal is armed with a kind of ivory sword, a halberd in the terminology of certain naturalists. This is a prin-

impractical (and probably painful); no narwhals have ever been found with scars or wounds from "jousts"; and spearing fish is highly unlikely because there would be no way for the narwhal to get at the fish impaled on its tooth. Female narwhals don't have a tusk (other than the tusk in males, narwhals have no erupted teeth at all), so the function of this preternaturally elongated tooth is one of nature's long-standing enigmas. The revelation that the tusk an extremely acute sense organ that enables the animal to measure minute changes in water temperature, pressure, and composition is remarkable, more so for its male chauvinist implications: because only the males have this extraordinary tooth, only they can sense their environment.

Examining a narwhal tooth with an electron microscope, Martin Nweeia of Harvard's School of Dental Medicine found that the tusk contained 10 million nerve endings that tunneled from the core to the surface and was therefore an incredible device for sensing its ambient surroundings. It is the only tooth known to have tubules that connect with the surface, which is more astonishing since cold is one of the things that tubules are most sensitive to, as humans sometimes discover when diseased gums are exposed. In an interview in the *New York Times*, where his research was discussed, Dr. Nweeia said, "Of all the places you'd think you'd want to do the most to insulate yourself from that outside environment, this guy has gone out of his way to open himself up to it." Swordfish generally avoid cold climates, although the water temperature a couple of thousand feet down is not exactly tropical. Could the sword also be a sensory device? In answer to my questions, Spanish teuthologist Angel Guerra wrote to me on the very day (December 13, 2005) that the news broke of the sensitive narwhal tusk: "I have carried out some preliminary observations in the bill of the swordfish. It is quite innervate [supplied with nerves], and possibly it has some sensorial systems (mechano- and/or chemo-

---

cipal tooth with the hardness of steel. Some of these teeth have been found embedded in the bodies of whales, which the narwhal always attacks with success." Of course there is no giant narwhal; even the small ones do not attack whales or ships; and the actual initiator of the attacks on ships was Captain Nemo at the helm of his lance-prowed submarine, *Nautilus*. (On p. **xx** of *this* book, you can read about a swordfish attack on the *Nautilus*.)

receptors). Therefore, in my opinion, the swordfish may not only use the bill to wound or kill prey, but also to detect them. This could be the explanation of why the swordfish is able to prey upon squid species that do not form schools, like *Architeuthis, Thysanoteuthis rhombus*, etc."

In a letter to me regarding the December 2005 revelations about the narwhal tusk, Harry Fierstine wrote, "In my own experience, i.e., sectioning bills at different levels and looking at them under a microscope, I have never seen any canals leading to the periphery of the rostrum, and I have looked for them. Of course, maybe if I had prepared sections for the electron microscope I would have seen them. Dentine has tubules, acellular bone does not, and the rostrum of swordfish and istiophorids is composed of acellular bone. However, the paired longitudinal canals in istiophorids (and presumably swordfish, too) contain nerves and blood vessels. In summary, there could be branches of the longitudinal canals leading to the periphery of the bill, but I doubt it."

Like swordfish, narwhals are also accomplished deep divers. When on their wintering grounds, narwhals make some of the deepest dives ever recorded for a marine mammal, diving to at least 800 meters (2,625 feet) over 15 times per day, with dives that can reach 1,500 meters (4,921 feet). Dives to these depths last around 25 minutes, including the time spent at the bottom and the transit down and back from the surface. Narwhals have a relatively restricted and specialized diet, consisting mostly of Greenland halibut, Arctic cod, polar cod, squid and shrimp. Other stomach contents have included wolfish, capelin, and sometimes rocks, accidentally ingested when the little whales feed near the bottom. (The Greenland halibut [*Reinhardtius hippoglossoides*] is a flatfish, classified with the Pleuronectidae, or right-eye flounders. It is more than a little difficult to envision a narwhal, with a lengthy spike getting in the way, feeding on a bottom-dwelling flounder.)

There are abundant differences between the narwhal and the swordfish. The narwhal is a mammal, with a large and complex brain and the ability to locate prey by echolocation; the swordfish has a much smaller brain (but much bigger eyes) and probably does its hunting by sight. Both animals are about the same size, both have a prominent protrusion in the anterior position, and the "reason" for this protrusion in both cases—if we really need a reason—is still a mystery. The comparison between the two deep divers is further complicated by their pronounced sexual dimorphism: only the male narwhals have the tusk, and only the female swordfish reach mammoth size.

Patricia Ribeiro Simões and José Pedro Andrade of the University of Algarve in Portugal found that male and female swordfish have different diets. Analyzing the stomach contents of 58 females and 24 males taken in the Azorean longline fishery, the researchers identified 11 species of fish and 15 cephalopod species. Fish common to both sexes were the cutlassfish (*Lepidopus caudatus*), the chub mackerel (*Scomber japonicus*), the horse mackerel (*Trachurus picturatus*), and the boarfish (*Capros aper*). The cutlassfish is a large aggressive species, up to seven feet in length, that was predominantly found in the stomachs of females, while the boarfish, which gets to be only about six inches long and lives in schools near the surface, is the favorite fish of male swordfish in Azorean waters. The smaller fishes, such as lantern fishes (Myctophidae) showed up mainly in the stomachs of males. The same thing was evident with cephalopods: females ate the larger species (such as *Ommastrephes bartrami*, which attains a mantle length of 20 inches), while male swordfish preferred *Histioteuthis*, a photophore-studded species that gets to be about twelve inches long. Because different fish and cephalopod species occupy different habitats and different depths, we can assume that male and female swordfish do the same (at least for feeding), and we can therefore add this idiosyncracy to the catalog of swordfish mysteries.

Very few of the studies mentioned sharks as prey items of swordfish, even though many species are small enough to pass down the swordfish's gullet, and there are even some—the lanternsharks, for example—that bioluminesce *and* aggregate, which should make them very likely targets. Maksimov reported on the remains of a carcharhinid shark found in the stomach of a 376-pound swordfish caught in the Gulf of Guinea in 1965 that "showed no sign of a blow from the sword. The position of the shark in the stomach of the swordfish indicated that it had been swallowed head-first, a mode of feeding characteristic of all tuna-like fishes." But, he wrote, "despite the profusion of different shark species . . . no shark remains were found in any of the other 500 swordfish stomachs examined from the South Atlantic."

We know that mako sharks attack and eat swordfish, but there are little sharks of the genus *Isistius* that attack not only swordfishes but also marlins, dolphins, pinnipeds, and even large whales by taking little circular bites out of them—which accounts for their common name, cookie-cutter sharks. *Isistius brasiliensis* is a 20-inch-long shark with small erect teeth in the upper jaw and large triangular teeth in the lower, which affixes itself to its prey with its suctorial lips and then spins to cut out a cookie-shaped plug of flesh from the larger animal. The source of these "crater wounds" was long a mystery, but biologist E. C. Jones solved the mystery by holding the open mouth of a dead *Isistius* up to a nectarine and twisting it as he imagined the shark would do as it bit a larger animal. The result was perfect pluglike core removed from the fruit, exactly like the piece that would have been removed from the living victim. (It evidently preys—or attempts to—on nonliving victims as well, as evidenced by the appearance of the same bites on the rubber coating of the sonar domes of submarines.) The cookie-cutter is brightly luminescent and probably attracts predators with its lights and then turns the tables and takes a bite out of its attacker. These "attacks" on swordfish are so common that in a study of crater wounds on swordfish undertaken by the Spanish longline fleet in the northeastern Atlantic, Spanish biologists Muñoz-Chápuli, Rey Salgado, and de la Serna were able to determine the northeastern Atlantic distribution of the little shark.

In a review of the functional morphology of the swordfish, Soviet ichthyologist V. V. Ovchinnikov dismissed the notion that the fish used its sword as an offensive or defensive weapon and wrote,

> The view that the sword is used for defense purposes and for attacking large animals is implausible. Fragments of rostrum are often found in the stomachs of Galeocerdo [tiger] sharks and Balaenoptera whales, sometimes even in xiphioids themselves. It would be expected that the swordfish's speed would be enough of a protection against its enemies, since even sharks do not move as swiftly as the swordfish. An attack on finback whales could hardly be in defense since these animals feed on small fish and do not represent any danger to the swordfish. Nor does the latter, as evident from the structure of its oral apparatus, feed on whale meat.

Even though we don't see swordfish in action very often—except perhaps when they are "sunning" (which does not entail much action) or fighting an angler (which does)—we know something about their activities, largely because of telemetric tags that have been affixed to individual fishes, enabling scientists to track their horizontal and vertical movements. The earliest experiments were conducted by Carey and Robison, who attached transmitters to seven swordfish, in the Pacific off Cabo San Lucas and in the Atlantic off Cape Hatteras.[4] The movements of the fish were tracked by hydrophones, and it was observed that they dove deeply during daylight hours and came to the surface at night. The authors commented,

> There is an obvious relationship between the vertical movements of swordfish and light. The most rapid changes in depth were during a 2-hour period at dawn and dusk when surface illumination changes by six or seven orders of magnitude, and the greatest depth was reached at noon, when light at the surface was at a maximum. . . . The swordfish also appeared to respond to moonlight. [When there was no moon] the swordfish were usually at depths <10m and often right on the surface. . . . In other experiments there was a full moon shining [and] the fish were swimming at a greater depth in response to moonlight, although the wind might also have had an influence.

Nowadays, the transmitter/hydrophone system has been replaced by more sophisticated sensors that are implanted in the skin and record

4. Over a long and distinguished career, Francis G. Carey (1931–94) studied the physiology, functional anatomy, and behavioral ecology of large fishes, including the mackerel sharks, blue sharks; bluefin and other tunas, the swordfish, and the marlins. In 1966 (with John Teal) he identified warm-bloodedness in tuna, and by 1973, he incorporated the mackerel sharks and the skipjack and other tunas, as well as the wahoo, into the category of warm-blooded fishes, a previously unsuspected category. In the early 1980s he put a temperature probe in the muscle of a great white and found that the shark's body temperature was higher than the surrounding water. A few years later, John McCosker, an ichthyologist at the California Academy of Sciences, induced a great white to swallow a temperature probe buried in a slab of seal blubber and confirmed Carey's discovery that the great white shark is warm-blooded.

the depth, swimming speed, and temperature of the fish. In a 1998 study, Barbara Block and her colleagues tracked bluefin tuna that were outfitted with "pop-off satellite tags" that recorded the movements of the fish and also the water temperatures; the data were reported to computers via satellite. (But you still have to catch the fish in order to attach the tag.)

In a 1990 study, Carey repeated the telemetry experiments of 1981, this time with swordfish in the New York Bight (an indentation in the Atlantic Coast), the Straits of Florida, and Georges Bank (an area of elevated sea floor between Cape Cod and Cape Sable Island, Nova Scotia). The results were similar, strongly suggesting that swordfish around North America (and elsewhere by extension) behaved pretty much the same way. Of "the activities of swordfish," he wrote:

> These activities included a daily cycle of vertical movements in which the swordfish were on the surface at night but went to depths as great as 600 meters [2000 feet] during the day. The fish experienced temperature changes as great as 19°C in these vertical movements. . . . Some of the swordfish had daily cycles which took them to the bottom on an inshore bank during the day and out over deep water on the surface at night.

Even though scientists may question some of their observations, anglers have always known that swordfish and marlins strike at their prey with their bills before consuming it. When a bait is being trolled behind the boat, the fish first hits the line or the bait with its bill. Zane Grey: "I learned to recognize the sharp vibration of my line when a swordfish rapped the bait with his sword. No doubt he thought he thus killed his prey. Then the strike would come inevitably soon after." Grey's "swordfish" in this case was a marlin, but when Eugenie Marron described the hit on the line, she was talking about what she called albacora, the broadbill:

> The albacora whams out with his bill. "Strike," the fisherman shouts and quickly throws his loop into the sea. Now the angler sits at attention, watching the reel and waiting for the big fish to pick up the bait. Again the albacora whams, but he is more a killer than a feeder. The boat is out of gear, just sitting on the surface, waiting, and the

albacora lashes out at the bait, toying with it, or torturing it if you will, . . . Since albacora would rather kill than feed, it is almost impossible to hook one in the mouth. When the albacora whams at the bait with his bill, hooks often snag him in the fins or the tail.

Swordfish are not schooling fish, and it is rare to see more than one at a time. Before longlines, fishermen would hunt them at the surface, sending a lookout up the mast to scan the waters for the telltale fins breaking the surface. The two sickle shapes of the dorsal fin and the upper lobe of the tailfin would indicate that the fish was lolling just below the surface, moving at a leisurely pace. This behavior was called finning, which is self-explanatory, and also sunning, which would appear to be a way for the swordfish to warm its body or replenish its oxygen supply. For the most part, swordfish dive deeply during the day, but they evidently spend some time basking in the sun. Their ability to float on or near the surface, says Carey, is a function of the swim bladder:

Swordfish swimming on the surface seem to have neutral or sufficient positive buoyancy to raise the dorsal and caudal fins out of the water. Swordfish taken on longlines frequently float, swim bladders distended, when hauled to the surface, and would have been at neutral buoyancy at a pressure of a few atmospheres. The swordfish are clearly able to inflate their bladder to a volume which will give them neutral density at some near surface depth.

When we hear the phrase "deep-sea fishes," we tend to think of small, bioluminescent creatures or, perhaps, the lumpy deep-sea anglers, with a lighted lure to attract prey fishes in the blackness of the depths. Fishes that live at great depths, such as eelpouts, grenadiers, or ratfishes, rarely reach a length of two feet, so when Ron Church took the submersible *Deepstar 4000* to its eponymous working depth of 4,000 feet in the Gulf of Mexico in 1971, he expected to see—and photograph for a *National Geographic* article—a lot of strange little fishes, along with weird sponges, squids, crabs, and lobsters. What he didn't expect to find at those depths was a swordfish. He found not just one, but a lot of them. "Out of the nineteen dives in and around the De Soto Canyon," he wrote, "broadbill were sighted on nine of them. On every dive within 5 miles of this location at least one broadbill was sighted." He continued: "While busily watching small hatchetfish and lantern fish, which seldom get larger than 3 inches, it is almost shocking to have a streamlined silver giant nearly 50 times their size suddenly materialize into view." One materialized outside the viewing port of the submersible at 2,145 feet and Church's photograph was used to illustrate his 1968 article in *Sea Frontiers*, "Broadbill Swordfish in Deep Water." That swordfishes spend so much time in deep water—especially during daylight hours—might go a long way to explaining why they are so difficult for fishermen to *find*, let alone *catch* at the surface.

When they descend to depths of 2,000 feet, swordfish experience large and abrupt temperature changes that would chill the brain and effect the central nervous system, so they have developed a heater that

Figure 3.5. In pursuit of their prey, swordfish range from the sunlit surface waters to the blackness of the depths. Painting by Richard Ellis.

keeps their brain and eyes significantly warmer than the surrounding water. In a study published in 1982, Carey identified the heat-exchange mechanism in the brain of the swordfish, which doesn't necessarily make them any smarter than other fish but only more successful at what they do and where they do it. Carey also identified a corresponding structure in the brains of marlin and spearfish, but these Istiophorids are "thought to be surface dwellers of warm tropical seas" and, therefore, "the utility of a heating system for the brain is not apparent in this environment, but . . . it can only be assumed that there are situations where it is of advantage to them.") Heating the eyes may also prove advantageous to a fish that hunts in semidarkness, but as with so much in the biology of the billfishes, the eye- and brain-heater is unique and serves to distinguish them from other fishes. According to Barbara Block:

> What truly makes billfishes unique, in terms of their thermal strategy, among not only the teleost fish but within the entire animal kingdom, is the presence of a mass of heat producing (thermogenic) tissue in the head which warms the brain and eyes. The specialized thermogenic organ acts as a furnace and keeps the brain and eye temperatures significantly above that of the ambient water temperature. The heater organ supplies both the brain and eye with warm blood, and provides all of the billfishes with the physiological capability of regulating their brain and eye temperatures.

Swordfish dive deep at the first glow in the morning and rise at dusk as the sun begins to set. Their huge eyes, as big as grapefruit, are so sensitive that the fish can sense moonlight. Although they occasionally bask at the surface in full sun, they prefer to dwell in semidarkness. While the head of the swordfish is kept warm by thermogenic tissue, it heats the rest of its body by basking in the sun on the surface. When it descends to hunt, its body cools slowly. The larger the fish, the longer it retains heat; Carey thinks that a 600-pound swordfish could stay below the thermocline for days. During these daily migrations, the swordfish encounters water temperature ranges of up to 22°C. Cold temperatures impair the nervous system, and one would expect these broad swings to render the swordfish stuporous, but in an article in the Woods Hole magazine *Oceanus*, Carey described the "temperature tricks" that the swordfish employed. Carey's studies showed that the swordfish's eye muscles and tiny brain ("about as big as the last joint of my index finger") are swaddled in a mass of brown tissue ("about the size of a hen's egg") the color and consistency of liver. This brown blanket is suffused with warm blood from its own large rete mirabile and is built from modified muscle cells that produce heat. The organ is 50 times larger than the swordfish brain and pumps out enough heat to raise the fish's cranial temperature to 14°C above the surrounding water. And swordfish spend a lot of time in deep, cold water.

The great majority of the 20,000 known fish species lose the heat generated by their metabolism through their gills and thus maintain a body temperature that is more or less the same as the water they are swimming in. They are known as cold-blooded fishes. In 1966, Frank Carey and John Teal of Woods Hole Oceanographic Institution observed that "in a tuna a highly developed countercurrent heat-exchange system located in the vascular system of the muscle provides a thermal barrier which prevents heated blood from being carried off and lost through the gills. The heat-exchange system also lowers the thermal gradient between the surface of the body and the water and reduces heat through surface cooling." The flesh of the tuna is predominantly red muscle, which, with its heavy concentration of mitochondria and myoglobin, provides energy for endurance activities. Bluefin tuna migrate for great distances and often at great speeds; one fish tagged off the Bahamas in June 1962 was captured off Norway 50 days later after a migration of 6,200 miles. The flesh of swordfish is mostly white muscle, which is designed for sudden bursts of activity. Swordfish generally move with prevailing currents, where they use their acute vision to locate their prey and their superior acceleration to chase it down. Unlike some tuna species and the mackerel sharks (great white, mako, and porbeagle), swordfish do not have the countercurrent system that warms their muscle tissue, but they do have a unique muscle tissue that warms the blood flowing to the brain and eyes when the fish is in cold water.

"Vision is obviously important to swordfish," write Carey and Robison, "The eyes of a 150 kg fish are as large as an orange and almost touch in the midplane of the skull." Just how important vision is would be discovered later, when researchers learned that swordfish see in color; they have the best vision of any fish; and they have a special device that heats their eyes. In a 2000 paper, Kerstin Fritsches of the Vision, Touch, and Hearing Research Center of the Physiology Department of the University of Queensland, along with Julian Partridge, John Pettigrew, and N. Justin Marshall, first demonstrated the possibility that marlins, sailfish, and swordfish might see in color. Earlier studies had suggested that these fishes were "monochromats" (animals that see in shades of a single color), but Fritsches and her colleagues presented evidence of two distinct cone types in the eyes of blue and black marlins and also sailfish, which does not necessarily prove that they can distinguish colors, but they certainly have the equipment. Then using something called microspectrophotometry, Fritsches and Warrant reported that "many fish have now been identified as [having] supreme color vision potential." The authors identified certain pigments in the eyes of deep-diving species such as bigeye tuna (*Thunnus obesus*) and the broadbill swordfish as being "perfectly matched to the prevailing light in clear blue, deeper ocean water, suggesting that all species tested have adapted their visual system to optimize light sensitivity at deeper diving depths."

Even before microspectrophotometry, however, Zane Grey knew that swordfish had exceptional vision. In *Tales of Fishes*, as part of his limitless celebration of the swordfish, he wrote, "a swordfish could see as far as the rays of light penetrated in whatever depth he swam. I have always suspected he had extraordinary eyesight . . . No fear swordfish will not see a bait. He can see the bait and the boat a long distance." Then Kerstin Fritsches, Richard Brill of National Marine Fisheries Service (NMFS) Laboratory in Washington, DC, and Eric Warrant of the University of Lund in Sweden proved that Zane Grey was right—though he didn't know the half of it. Their 2005 study showed that "warming the retina significantly improves temporal resolution, and hence the detection of rapid motion. . . . Depending on diving depth, temporal resolution can be ten times greater in these fishes than in fishes with eyes at the same temperature as the surrounding water." Swordfish are primarily visual predators, and heating their eyes enables them to pick out fast-moving prey animals in the reduced light levels of the depths. The authors concluded:

> Their cold-blooded prey . . . will have eyes at the same temperature as the surrounding water and thus, potentially much lower temporal resolution, diminishing their ability to visually avoid predation. Given the speed and maneuverability of the swordfish's cephalopod prey, such as the large flying squids of the family *Ommastrephidae*, the large, fast, and sensitive eyes of swordfishes give them a crucial advantage in pursuing and intercepting fast moving prey in the cold and dimly lit depths of the ocean.

The swordfish knows what to do at the depths, too. With the obvious exception of certain deep-sea fishes that spend their entire lives in the blackness of the depths, many species rise toward the surface at night to feed. Ever since we have been able to send sounds into the ocean and analyze the returning echoes, we have noticed an element that moves through the water column like a variable bottom. The existence of this phantom, sound-reflecting layer was not officially recognized until 1942, when Navy scientists, experimenting with underwater sound for detecting submarines, regularly encountered an echo from around 900 feet where no bottom existed; the water was actually several thousand feet deep. They knew it was not a submarine because it was diffuse, rather like a moving shadow, and most curiously, it appeared only during the day. By evening it would appear to rise toward the surface and disperse. Even before the nature of this layer was identified, it was christened the deep-scattering layer, quickly abbreviated to DSL.

The swim bladder in fishes is a gas-filled organ that acts as a hydrostatic device; by adjusting to variations in pressure, the fish can remain poised at any depth without rising or falling. This equilibrium is achieved by making the density of the fish nearly equal to that of the surrounding water. Net hauls have shown that lantern fishes migrate vertically, rising toward the surface at night and descending during the daylight hours. As this movement coincides with the behavior of the DSL, it is not difficult to attribute these heretofore mysterious echoes to the lantern fishes. Lantern fishes exist in huge numbers of many different kinds; there are more than 240 species in 30 genera. They are all small, silvery fishes, ranging in size from two to six inches, with a large head and large eyes. These little fishes have 50–80 photophores on the head, belly, and sides, which emit a startlingly bright blue light, likened to an electric spark. Using a photometer to measure penetration of light into the sea and also the flashing of luminescent animals, Clarke and Backus demonstrated that these flashes were closely related to the vertical migration of animals in the DSL. They found that the "flashiest" of these migrators were the lantern fishes. Furthermore, in order to qualify as a major component of the DSL, the sound-scatterers would have to be nearly worldwide in distribution, and the myctophids fill this requirement quite nicely. They are found throughout the worlds' oceans, but not in the high latitudes of the Antarctic, and, perhaps not coincidentally, the DSL is not encountered there either. And neither are swordfish.

The lantern fishes and other creatures of the DSL move through the water column in order to maintain the luminescence that (might) conceal them from predation. During the daylight hours they hide in the depths, but as night approaches they begin to rise toward the surface where their photophores will allow them to maintain the integrity of the massed aggregations, attract a mate, and feed on the zooplanktonic creatures of the pelagic zone. Of course not everything about bioluminescence is beneficial to the lantern fishes; the flashing lights also attract predators. As we have seen, swordfish feed on lantern fishes, so it is possible—indeed, likely—that the swordfish's nighttime dives constitute a rendezvous with the rising components of the DSL. Affixing telemetry devices to swordfishes in deep water, Frank Carey, along with his colleagues, found that "where there is a well-developed community of vertically migrating animals . . . the swordfish moves in the same manner, and may locate itself in the most dense region of the scattering layer." Spotting the tiny blue lights with its enormous, light-gathering eyes, the great fish darts into the densely packed school, and wielding its sword as if it were—well, as if it were a *sword*—it slashes and stuns myctophids by the hundreds, gobbling them up as they float toward the bottom. For the daylight hours, it has developed a different strategy, herding mackerel, menhaden, bonitos, and other schooling surface fishes against the barrier of the surface itself and, then, employing the daytime version of the slash-and-gobble method.

Along with the bluefin tuna, swordfish are among the few large pelagic predatory fish that function efficiently at the surface and near the seafloor. They are adapted for life in the frigid blackness of the depths and in the sun-warmed green waters of the surface. There are some shark species that inhabit deep water, and others that live and hunt close to the surface, but none that are as versatile as *Xiphias*. (There are some records of deep-diving white sharks, but for the most part, *Carcharodon carcharias* feeds near the surface, most frequently on seals and sea lions.) We might learn more about the swordfish if we could observe one in captivity, but so far, even the *idea* of such a thing is impracticable.

Aquarists have maintained almost everything else, from whale sharks, white sharks, sawfish, and even bluefin tuna, in tanks for public observation. The million-gallon Outer Bay tank at the Monterey Bay Aquarium in California is home to dozens of large bluefin and yellowfin tuna, and from September 2004 to October 2011, six juvenile great white sharks were exhibited in the Outer Bay tank. (The first was on exhibit for six months; the second for four months; the third for five months; the fourth for 11 days; the fifth for 69 days; and the sixth, released in October 2011, for 55 days. All were tagged for research purposes and successfully returned to the wild, except the last one, which died a week after it had been released.) There have been only two large whales ever held in captivity; they were gray whales, both of which were captured as juveniles, both of which were kept at Sea World in San Diego, and were released when they got too big for any tank and too expensive to feed. To date, there have been no successful attempts to capture a coelacanth; most large squid are too excitable to be kept in captivity—they tend to jump out of the tanks—and the idea of exhibiting a billfish is almost oxymoronic. You can't capture one except by fishing for it, a process that inevitably injures the fish, and even if you could somehow get one into a tank, their temperament and armament seem designed for tank- or self-destruction.

4 Armed and Dangerous

Swordfish steaks are delicious and the fish has been sought as food by the inhabitants of the northern Mediterranean since before the dawn of history. Fishermen in boats and armed with darts, tridents, or harpoons, have hunted the swordfish in this sea, and particularly in the Tyrrhenian subdivision, from far antiquity down to the present. And in all these long years their boats have been subject to retaliatory attacks by the fish.
—E. W. Gudger, *On the Alleged Pugnacity of the Swordfish*

On April 15, 2003, Mark Ferrari was swimming in water 250 feet deep in the AuAu Channel between the Hawaiian islands of Maui and Lanai. A whale researcher and filmmaker, Ferrari had been working for years in these waters with his wife and fellow researcher, Debbie Glockner-Ferrari. Their primary interest was the behavior of humpback whales, but on this day, Mark was following a school of false killer whales (*Pseudorca crassidens*), rarely photographed toothed whales that are akin to pilot whales and killer whales. (Killer whales, pilot whales, and false killer whales are actually large dolphins.) There were 50 or 60 of the coal-black whales, including many mothers and calves. Mark had his underwater video camera, but he was without scuba gear because he believed that the bubbles disturbed the whales, and besides, most whale action takes place close to the surface, where he and the whales could get a breath of air when they needed it. The false killers, some of which were more than 20 feet long, swam rapidly in and out of his view, sometimes coming disarmingly close and "smiling"—baring their teeth—before swooping away with powerful beats of their horizontal tail flukes. In the clear blue water, off to his right about 30 feet away, he spotted a shadow, not quite as large as the false killers, but of a completely different configuration. From its protruding snout and vertical tail fin, he recognized it as a billfish, about 15 feet long. Because female billfish are larger than males, and because this was a *very* large specimen, Mark assumed it was a female. Was it a swordfish? A marlin? He shot a couple of frames of the fish, which seemed to be hovering in the water about 20 feet away, and when he lost sight of the false killers and the fish, he thought the encounter was over and climbed back in the boat.

At the surface, the black false killers with their curiously humped pectoral fins were leaping excitedly, and Mark decided to reenter the water to see what was going on. Big mistake. The *Pseudorcas* had formed a sort of net beneath the billfish, and every time she tried to escape by diving, one of them rushed in and bit a chunk from her flank. Mark was now ten feet from the wounded fish, and for a reason that will never be known, she charged directly at him. Was it mindless panic? A defensive maneuver aimed at whatever she could hit? He was

Figure 4.1. Seconds after Mark Ferrari took this photograph of a swordfish in the waters off Maui, the fish turned and charged him, stabbing him through the chest and nearly killing him. Photograph by Mark Ferrari.

struck high on his right chest, just at the base of the neck; his clavicle was broken and his scapula shattered. The sword, some of which had broken off in Mark's body, had missed his carotid artery by less than half an inch; a little lower and it would have punctured his lung. The sword did not pass entirely through him, and with a toss of her head, the fish disengaged and swam off. Bleeding badly, Mark surfaced and called for Debbie, who was in the boat not far away. Once ashore they loaded him into an ambulance, held wet towels over the gaping wound, and drove him across the island to Maui Memorial Hospital in Wailuku. In addition to the broken bones, there was extensive nerve and muscle damage. He had lost so much blood that the surgeons thought he might not make it.

In his book about the North Atlantic swordfishery, Charles Dana Gibson (ex-commercial swordfisherman and grandson of the artist of the same name) wrote, "The question of whether an unprovoked swordfish would attack a free-swimming man has often been debated. To date, there has not been one reported case, but this is far from conclusive when one considers how rare it would be for one to be swimming in waters frequented by swordfish." It might be argued that the fish that attacked Mark was "provoked," but certainly not by him, and therefore, Gibson's statement is no longer true. Mark Ferrari lived to tell his story (he told it to me in December 2004), but in 1886, Captain F. D. Lansford, a commercial fisherman, was killed fighting a hooked swordfish. The swordfish charged Lansford's dory, causing the line to go slack and the captain to fall backward into the boat; the fish then rammed the boat, stabbing through both boat and man. The dory with the swordfish attached was hauled aboard the schooner, the sword cut off and the fish killed. Lansford died three days later (Wilcox 1887).

Fishing (and making a film) off Venezuela's La Guaira Bank in 1994, artist and scientist Guy Harvey jumped in the water with a hooked swordfish. In his book *Portraits from the Deep*, he recounts the experience:

> As I approached I began picking out the details. The hook was in the right corner of the jaw; the bill was so long I had to turn the camera and zoom back to 20mm to fit it all in. The fish started to kick harder as I got a final shot looking along its body. Then it pulled

away strongly, swam up a few more yards and suddenly turned around and charged right at me. I was mesmerized—frozen to the spot, totally focussed on what was taking place! With great speed the fish approached, but then it stopped abruptly and turned sideways, looking at me with its huge, featurelss black eye. I exhaled a blast of bubbles as it turned away and headed up to the surface, where it began jumping.

Undoubtedly because of its size and armament, the swordfish has always been regarded as a dangerous creature, and probably always will be. Here for example, is the description of "*Xyphias gladius communis*," from William Dewhurst's 1835 *Natural History of the Order Cetacea and the Oceanic Inhabitants of the Arctic Regions*: "The common sword-fish is a native of the Mediterranean and Sicilian Seas; it grows to a very large size, sometimes measuring 20 feet in length; it is active and predacious, feeding on all kinds of fishes, and it likely a formidable enemy to the whale, which it destroys by piercing it with its sword-shaped snout." Never mind about mercury—which may be at higher levels in swordfish than any other food fish—this is a fish that can kill you.

Among the documented "victims" of swordfish attacks are sea turtles, mako sharks, tuna, and whales. In the hands of a man, a sword might be used to chop things up, but on the end of a fish's nose, chopping doesn't seem like a reasonable option, so we have to look elsewhere for the "reason" a swordfish might stab a whale or a turtle. In a discussion of sea turtles stabbed by swordfish, Harry Fierstine, who has made a career of studying swordfish, past and present, wrote,

> An extended debate has centered on the reason why billfish stab large objects which they cannot eat. Since billfish swallow their prey whole, it is difficult to understand how even the largest and hungriest of these fish could attempt to eat an adult-size marine turtle, especially when viewing the chelonian from perpendicular to the plastron or carapace. There is no evidence of turtle of any size occurring in the diet of any billfishes.

There is a comparable lack of evidence for whales occurring in the billfish's diet, and yet in 1933, Australian marine biologist David Stead related the story of the crew of a fishing boat out of Auckland, New Zealand, who observed a great commotion on the surface, and as they approached it, they "discovered a cow whale [species unknown] defending her calf against the furious onslaught of a swordfish. After a long struggle the whale dealt a lucky blow when she gave the swordfish a tremendous stroke with one of her tailflukes, breaking the swordfish's large dorsal fin and apparently paralyzing it. The fishermen killed the disabled monster and brought it to Auckland, where it was found to measure 12 feet 6 inches in length." To a great extent, Stead relied for his natural history on fishermen's tales, so there is the outside possibility that the event didn't happen exactly as it was described to him. (Stead is also responsible for the story—reported in his *Sharks and Rays of Australian Seas*—of a ghostly white shark that was 115 feet long, spotted by fishermen in the waters of New South Wales in 1918.)

In an article about diving with blue marlins, ichthyo-artist Stanley Meltzoff attributed the stabbing of boats and flotsam by billfishes to be "only evidence that the marlin can't see straight forward or jam on his brakes. These prove bad driving at the surface rather than bad temper, but men facing a charging marlin cannot but think otherwise." Others have suggested that the billfish charging into a school of prey fishes might overshoot the mark and plunge its sword into the object under which the fish have taken refuge. While it is true that a torpedo-shaped billfish, moving through the water at speed, cannot easily "jam on the brakes" and come to a complete stop, the billfish has to be maneuverable enough to avoid such a collision by modifying its angle of attack, and if it misses the fishes, it ought to be able to turn and avoid the boat, or whale, or whatever. The inability to do this would certainly be maladaptive, especially as a swordfish with its bill stuck into a boat probably dies in the process of becoming unstuck. The attacks on dories by hooked billfish, the attack on Mark Ferrari, and the attacks on submersibles require better explanations than overshooting the mark. Even though swordfish attacks on whales are almost impossible to understand, the swordfish and the whale go back a long way together—in fact and fiction.

In Samuel Purchas's 1625 *Hakluytus Posthumus; or, Purchas His Pilgrimes, a Continuation of Hakluyt's Voyages*, there is a description of a cooperative attack on a whale by a swordfish and a thresher shark in the West Indies. According to Purchas's account, "Whilest we remayned at this Iland we saw a Whale chased by a Thresher and a Swordfish: they fought for the space of two houres, we might see the Thresher with his flayle lay on the monstrous blowes which was strange to behold: in the end these two fishes brought the Whale to her end." But what could that end be? Neither the thresher nor the swordfish have the dental equipment to eat whale meat, so Purchas's account seems like a case of multiple mistaken identities—or plain old fantasy.

We'll leave the final word on the story of the swordfish and the whale to Frank Bullen, sometime whaleman and full-time fantasist:

> Unlike most of the sea people who hunt mostly for food, and that obtained keep peace, the Sword-fish longs, apparently with irresistable desire, for some foe worthy of his attack. This it is that impels him to launch himself like some living torpedo at the vast bulk of the whale, for the sheer savage delight of stabbing viciously again and yet again at the living mass before him, even if, as in the case of the enormous bowhead, with its two feet thickness of blubber, he cannot draw blood. To bear away, impaled by one swift and powerful blow, the newly born calf from beneath its mother's protective arm is a supreme delight, and one that yields solid results in the shape of food, for the tender body of the young whale is easily divided by repeated blows of the sword.

Of course! The sword is used to chop up whales! Swordfish do not inhabit the icy, Arctic habitat of the bowhead, but this doesn't faze Bullen, for he saves the best story for the sperm whale:

> But this fiercely aggressive spirit has its drawbacks too. As, for instance, when the *Xiphias* attacks a sperm whale, and meeting the impervious mass of the head, rebounds helplessly, to be caught before recovery between the huge mammal's lethal jaws and devoured.

In the old seamans' tale about the relationship between the thresher, the swordfish, and the whale, the thresher would circle the hapless whale, slapping at the water with its sickle-like tail and beating the surface of the ocean into a froth to further confuse the victim. While the whale was thus distracted, the swordfish would pierce the defenseless whale in a vital spot and kill it. This cooperative venture was apparently undertaken because neither the thresher nor the swordfish could handle the large whale separately. Unfortunately, neither the swordfish nor the thresher includes whale meat in its regular diet, so this whole operation would seem rather unnecessary, if not downright wasteful.

The fable does point up an interesting comparison between the two alleged conspirators: both have an abnormally elongated appendage, the swordfish in the forward position, and the thresher in the aft. We have an idea of the uses the swordfish makes of its elongated snout, but the thresher's tail is a little more problematical. Generally, the tail is about equal in length to the body of the fish, and since a thresher can achieve a total length of 20 feet, that means ten feet of shark body and ten feet of tail. Among the uses suggested for this tail are the following, in descending order of credibility: it is used to round up fish

Figure 4.2. The thresher shark, often implicated with the swordfish in coordinated attacks on whales. As neither the swordfish nor the shark have teeth, the stories would appear to be apocryphal. Painting by Richard Ellis.

into a compact school for the shark to feed on; it is used to flip fish out of the water and into its mouth; it is used to slap seabirds sitting on the surface, so the shark can then gobble them up. There are "eye-witness" reports of these actions, and somewhat less well-documented stories about truly bizarre uses to which the tail has been put. One story, quoted by someone who said the story was told to him, tells of a longline fisherman who had hooked a large shark, and when he leaned over the gunwale, the tail of a thresher (for that is what it was) came whipping out of the water and decapitated him.

What are we to make of this eyewitness account, written by P. F. Major and published in the otherwise reputable *Scientific Reports of the Whales Research Institute of Tokyo*?

> Splashing, boils of water and the backs of large whales breaking the surface were initially seen. The first evidence that billfish were involved, was the rolling to its right of a whale, such that a billfish, firmly embedded in the whale's left flank, was lifted clear of the water to a near vertical position above the whale. The billfish thrashed back and forth, snapped off its bill and fell into the water as the whale turned upright and submerged. Minutes later another billfish was lifted out of the water to a position about one-half to one-quarter from the vertical above a whale. The whale rolled back into the water with the billfish still embedded. Splashing, water boils and the backs of whales were seen for an additional few minutes, and then abruptly stopped.

It appears that at least two marlins were attacking some unidentified whales that were objecting to being bayoneted, but other than that, we have no idea what was going on.

In 1959, the Norwegian cetologist Åge Jonsgård wrote of two blue whales that were being flensed aboard Antarctic factory ships, which were found with swordfish swords embedded in the blubber. Jonsgård asked "whether our material can in any way throw light on the question what happened when the swordfishes stabbed the blue whales?" He cites a discussion in Norman and Fraser's 1938 book, wherein thresher sharks and swordfish were attacking an "enormous whale," but discounts it as a sea story. In 1923, Charles Haskins Townsend submitted a delightful little story to the *Bulletin* of the New York Zoological

Society, which he called "The Swordfish and Thresher Shark Delusion." Townsend felt that "with a million cameras making records of what happens in the animal world, the persistence of old-time misconceptions in natural history is a fact naturalists have to reckon with." In this case, the "misconception" was that the swordfish and the thresher shark sometimes attack whales. After quoting several old-time tales, Townsend (who was the director of the New York Zoological Society's aquarium) attributes the "attacks" to killer whales, where observers confused the orca's tall dorsal fin with the long tail of the thresher, and he suggests that the swordfish might strike the bottom of a ship (or a whale) as it charged at fishes that had taken shelter under the larger object.

Then in 1962, Jonsgård reported on three fin whales, also taken aboard Antarctic factory ships for flensing, with a broken-off sword in each one. With these five records, more or less from randomly examined whales, Jonsgård suggests that "it seems possible that whales are stabbed by swordfish to a far greater extent than we have hitherto reckoned. This possibility is supported by the accounts contained in literature of the fights between swordfish and whales, which state that the whales are stabbed by the swordfish from the underside. No whales have so far been found which were stabbed from below, but this does not in any way signify that this is unusual. It must be supposed that a whale which is stabbed in the belly has very small chances of surviving."

Why would a fish-eating swordfish stab a whale in the first place? The evidence from the whale carcasses indicates that the sword was broken off in the whale's flesh or blubber, which would mean that the fish—even it survived the wrenching off of its sword—would be unable in future to use the sword, whatever its use might be. Stabbing whales, turtles, boats, bales of rubber, and submarines suggests something unheard of in the animal kingdom: random aggression that results in injury and even death to the aggressor. (Bee stings sometimes result in the death of the bee because the stinger is pulled out, but it is believed that the bee is acting defensively not preemptively.) British Museum of Natural History ichthyologist Albert Günther did not doubt that swordfishes attacked whales, although he said "the cause that excites them to such attacks is unknown." He noted that "they follow this instinct so blindly that they not rarely attack boats or large vessels in a similar manner, evidently mistaking them for Cetaceans." Maybe the swordfish isn't the aggressor; maybe it feels threatened by the rubber, whale, turtle, or submarine and is responding self-defensively. There is also the possibility—also unique—that swordfishes sometimes do crazy, inexplicable things that could be inspired by raging hormones, a temporary mental breakdown, or phases of the moon. The "intensity of swordfish feeing activity increases around the full moons, particularly where the thermocline is deep," said Ward and Elscot in 2000, and in 2005, Portuguese scientists Manuel Neves dos Santos and Alexandria Garcia observed that the land-based longline fishery for swordfish showed an increase in catches during the full moon. Taking a baited hook is hardly a sign of derangement in a fish (although it is often tantamount to suicide), so swordfish are probably not exhibiting "lunatic" behavior when they increase their feeding activity during periods when the moon is full.

Were the swordfish who attacked items that were perceptibly inedible mindlessly aggressive or were there extenuating circumstances? The fish that hit Mark Ferrari was itself under attack by false killer whales, and the fish that killed Captain Lansford probably panicked as it was being dragged through the water by a thick hook through its mouth. Sometimes swordfish charge the boats from which they have been harpooned or hooked, but in other cases, such as attacks on whales, bales of rubber, or submarines, the attack comes out of the blue, as it were, with no identifiable motivating factors. Even sharks, long considered mindless man-eaters, do not attack people randomly or accidentally. Something inspires them to bite: it may be fear, hunger, territoriality, electrical stimulation, or confusion about what is edible or what is not. However, most people do not consider sharks malicious or evil but simply—as if this actually needed saying—animals that are profoundly different from us. They live in a world that Henry Beston described as "older and more complex than ours," and they are "gifted with extensions of the senses we have lost or never attained, living by voices we shall never hear." Most animal behavior studies do not involve attacks on people; but when tigers eat people, sharks kill surfers, or swordfish attack divers, we are forced to reconsider our attitude toward very big, very heavily armed animals.

In *Memoirs of the Royal Asiatic Society of Bengal*, Eugene W. Gudger published a paper titled "The Alleged Pugnacity of the Swordfish and the Spearfishes as Shown by Their Attacks on Vessels." Before he gets to the attacks, however, he introduces us to the putative attacker:

> Shaped on the lines of a mackerel, Xiphias has a head somewhat wedge-shaped with the upper jaw prolonged horizontally into double-edge sword. On the rounded back the non-depressible first dorsal fin has a fine backward rake. The second dorsal and the anal (placed under it) are very small, but the pectorals are large and powerful fins. The fusiform flexible body is thickest in the shoulder region. Whence it tapers gracefully back to a slender caudal peduncle strengthened on each side by a horizontal keel. And ending it all is the huge bilaterally symmetrical lunate tail fin, the great locomotor organ of this living torpedo.

Gudger was inspired to begin his study in 1936 by the Michael Lerner–AMNH–Swordfish Expedition to Louisburg, Cape Breton Island. In this lengthy paper—at 100 pages, it is almost a small book—Gudger reviews every "attack" that he could find and even works out a formula for the force of the blow with relation to the known speed of the various billfishes. (Most of the attacks are attributed to swordfishes, but there are also numerous records of marlin bills embedded in ship's timbers.) Gudger, who was not associated with the Royal Society of Bengal but was an associate in ichthyology at the American Museum of Natural History (AMNH) in New York, did not believe any of the stories of swordfish attacks on whales, citing the negative evidence of whaling historians Frederick Debell Bennett, Robert Cushman Murphy, and Roy Chapman Andrews, none of whom ever recorded a sword in whale blubber, although between them, they probably saw thousands of whales cut up. It would be up to a modern whaling historian to provide irrefutable evidence of swordfish attacks on whales.

In his account of the attacks by swordfish on almost anything and everything, Gudger begins with an account of the ancient Greek geographer Strabo of an attack on a fishing boat by a swordfish in the Strait of Messina (between Sicily and the toe of Italy) and then moves on to Pliny, Aelian, and Oppian, the Greek poet in whose *Halieutica* we learn that those who pursued *Xiphias* did so in "swordfish-shaped boats." (Because these swordfish-boatmen were "commercial" fishers,

Figure 4.3. "Fight between a Swordfish and a Whale." In this 1872 illustration, a very large swordfish attacks a very unusual whale, whose mouth is upside down and whose blowholes are spouting in opposite directions. Plate 25 from *The Bottom of the Sea*, by Leon Sonrel. Courtesy of the University of Washington Library.

we will meet them again.) Aelian was a Roman writer, born around 170 AD, who collected animal lore from Greek writers and presented it in a 17-volume collection known as *De natura animalium* (published in a modern edition titled *Aelian on the Characteristics of Animals*). That he had never laid eyes on a swordfish didn't seem to bother him, but he was certainly prepared to accept stories about their reputation for aggressiveness:

> Sword-fish . . . are suited to their name, witness the fact that the rest of their body is soft and harmless to the touch, that their teeth do not appear curved and sharp, that there are no spines springing erect from their back, as in the case of dolphins, or from their tail, but; what surprises one to learn and to see is this: the jaw just below its nose, through which it breathes and through which the stream flows to the gills and falls out, is prolonged to a sharp point, is straight and increases gradually in length and in bulk; it grows also as the fish grows into a monster and resembles the beak of a trireme. And the Sword-fish makes straight for fishes, kills them, and then feeds on them, and with this same sword beats off the attacks of the largest seamonsters. No smith has forged this weapon which grows upon the fish, and Nature has made it sharp. And so when these Sword-fish have attained a considerable size they even attack

> ships. And there are some who boast that they have seen a Bithynian vessel drawn up on shore in order that the keel which was suffering from age might receive the necessary attention, and fixed to the keel they saw the head of a sword-fish. For the creature had planted the sword given it by Nature, in the vessel, and when it attempted to withdraw, the whole of its body was rent from the neck owing to the force of the ship's onrush, while the sword remained fixed just as it entered originally.

Gudger's catalog of swordfish attacks on ships, boats, dories, and dinghies is broken down by species and by geography ("Attacks in the Bay of Bengal," "Attacks on Fishing Schooners in New England Waters," and so on); the study also contains an analysis of the speed of an attacking swordfish (estimated at 30 mph, but "it seems that the marlins and sailfishes are even faster swimmers") and the striking force of the fish. Gudger also includes a detailed, illustrated atlas of the skeleton of the swordfish and spearfishes, which he entitles "The Skeletal Structures of the Xiphiiform Fishes Make the Blows Possible." In his 1843 *Natural History of the Fishes of Massachusetts*, Jerome van C. Smith described an "attack," in which the fish came out the loser:

> On a calm summer day during last summer, a pilot was leisurely rowing his little skiff along the glassy bosom of the gently swelling waves, he was suddenly roused from his seat by the plunge of a swordfish thrusting his long spear more than 3 ft up through the bottom of the slender bark; when the pilot, with the presence of mind for which that whole fraternity are distinguished, broke it off on a level with the floor, by the butt of an oar, before the submarine assassin had time to withdraw his fearfully offensive weapon.

Gudger says that some swordfish encounters were collisions with "ships in passage," but "the attacks were mainly on fishing vessels by harpooned fish." Most of these "attacks" occurred in the western North Atlantic, where harpoon fishing for swordfish has been extensively carried on. Gudger describes no fewer than seven cases of swordfishes charging dories. One of these accounts, from 1937, goes as follows:

> We were cruising around off Montauk, about July 12, hunting for swordfish when our lookout at the masthead sighted a fish right

> ahead. The owner of the yacht ran forward to the swordfish stand and harpooned the fish. But unfortunately the harpoon struck forward of the dorsal fin and near the head. It usually happens that when a swordfish is harpooned in or near the head it seems to go crazy and starts looking for something to attack. This fish came to the surface after the first plunge downward and started cutting circles around the boat. We went onto it again and a second harpoon was driven into it. Still the fish would not go down and I put out in a dory to play it. I hauled on the line from the keg till I got within about twenty feet of the fish. Then it suddenly turned and like a flash drove its sword through the dory. Fortunately the sword did not strike me, but that was just my good luck. After striking the dory, the fish thrashed about so hard that it almost threw me out of the dory and did break off its sword at a point just below where it went through the bottom of the dory.

In a brief note in 1956, James Leonard Brierley Smith (always known as J. L. B. Smith), the discoverer of the coelacanth, discussed the inclination of swordfish and marlins to charge inanimate objects and pierce them with their bayonets. Off South Africa, after World War II, quantities of bales of rubber were floating in the ocean after their carriers had been torpedoed. "It was soon noticed," states Smith, "that many bales contained the tips of spears of marlins, and it now appears that the longer a bale remains in the sea the more broken spears it is likely to contain, for those cast ashore in recent times have held as many as four." One of these bales was pierced by a marlin spear to a depth of 13 inches, suggesting an extraordinarily powerful charge. Smith says, "Marlins must deliberately charge floating or submerged objects intending to impale them, possibly to secure food, but possibly also from plain aggressiveness." Not much food in a bale of rubber, but there may be food *under* it. We find a possible explanation in *Smiths' Sea Fishes* (a 1986 revision of J. L. B. Smith's *Sea Fishes of Southern Africa* [1950] edited by his widow, Margaret Smith, and Phillip Heemstra):

> The unfortunate marlin were probably chasing the small fishes that congregate under such floating objects and accidentally rammed the rubber bale with such force that the bill was driven up to the head in the bale. If the marlin were not able to dislodge the heavy rubber

bale (assuming the bill was not broken off in the initial impact), they would soon fall prey to sharks, which are quick to take advantage of such disabled fishes.

Because of his discovery of the coelacanth, J. L. B. Smith (1897–1967) was, in his time, probably the world's most famous ichthyologist—indeed, he may have been the only ichthyologist many people had ever heard of. Just before he died, he published "a collection of selected articles" in a book titled *High Tide*. The story of the coelacanth, included in this collection, represents Professor Smith's major contribution to science, but this little book also contains stories of dangerous sharks, barracudas, rogue waves, sea snakes, and marlins, already infamous as rubber piercers. "No marlin angler," he wrote, "should ever forget how dangerous these fishes can be":

> There is one man who will never forget this—an African who lives south of Port Amelia in Mocambique. During 1956, he was fishing one night at sea and had a lantern alight in his boat. A big fish took his bait and as he stood up, the better to control his line, suddenly a marlin came shooting up out of the water straight at him. Before he realized what was happening, its spear had gone right into his chest. He was knocked over and barely managed to grab the edge of the boat. According to what he saw, the marlin was longer than he was. Eventually when he had all but given up, the spear came out and, of course, blood poured from the wound. He managed to pull himself up, plugged the terrible gash with some rags and, after a rest, actually managed to reach the shore. There he got help and was taken as rapidly as possible to a hospital. It seems incredible that anyone could survive such a wound and experience. Probably no European ever could, but after some months in hospital this African came out fairly well recovered and lived to fish again—but not for marlin!

Rarely is a billfish attack on another billfish reported, but surely these big, powerful, and heavily armed fishes must engage in occasional altercations with one another. While evidence of such encounters is extremely scarce, it does exist. In a 1997 paper, Harry Fierstine reported "a large Atlantic blue marlin with two rostral fragments embedded in its head . . . the first record of a fish with multiple wounds."

The marlin, caught by a sport fisherman off Algarve, Portugal, in 1993, was said to weigh 789.7 kilograms (1,741 pounds), but submission of the details to IGFA did not result in a world's record because the weight was not confirmed by anyone except the fisherman. The application to IGFA, however, was accompanied by two rostral fragments (broken-off tips), and it was these fragments—one that came from the nape of the marlin and the other from the lower jaw—that were examined by Fierstine. He concluded that one had come from a striped marlin (*Tetrapturus audax*) and the other from a longbill spearfish (*T. pfluegeri*). We have no way of knowing the circumstances under which a huge blue marlin—it would have been the world's record holder if the application had been accepted (see p. 134)—was "attacked" by two smaller billfishes, nor do we know when in its life it was attacked. Fierstine concludes, "The Atlantic blue marlin apparently remained healthy in spite of two wounds and grew to enormous size because no vital organs were pierced." As for *why* a striped marlin might attack its larger relative, we haven't a clue.

One of the more inelegant of swordfish impalements is the one

Figure 4.4. "Sherman's Lagoon" © Jim Toomey, reproduced with permission of King Features Syndicate.

recorded—evidently from first-hand observation—by David Starr Jordan and Barton W. Evermann in their turn-of-the-twentieth-century discussion of the fishes of North America. They wrote:

> Swordfish fight gamely on the surface or below when harpooned. Storer wrote long ago that they sometimes sound with such speed and force as to drive the sword into the bottom, which fishermen say is by no means uncommon; and we saw this off Halifax in August 1914, when a fish more than 10 feet long, which we had harpooned from the *Grampus*, plunged with such force that it buried itself in the mud beyond its eyes in 56 fathoms of water. When finally hauled up alongside it brought up enough mud plastered to its head to yield a good sample of the bottom.

Hard to imagine the mud not being washed off as the fish was hauled up through 56 fathoms (336 feet) of water, but maybe the mud of the Halifax depths is especially sticky. Gudger finds similar accounts of *Xiphias* driving its sword into the bottom, including this one from Conrad Gesner, whose *Historia animalium* was published in Zurich in 1558: "When *Xiphias* sees a whale, he is filled with so much terror that he drives his sword into the earth, or a stone, or anything else that he finds at the bottom of the shallow water. Thus fixed by his head, he holds himself still. The whale indeed takes him for a log of wood or something like thereto, and ignoring him, passes by."

England and the United States both capitalized on the underwater aggressiveness of the billfishes by naming submarines after them. Both the US and UK navies had crafts named *Swordfish* and *Spearfish*; the US also had submarines named *Marlin* and *Billfish*. But in his novel *Twenty Thousand Leagues under the Sea*, when Jules Verne needed a name for the submarine that actually attacked ships with its iron prow, he chose *Nautilus*, which means sailor in Greek, but is the common (and scientific) name of a group of cephalopods with dramatically striped shells, whose behavior is the very definition of unaggressive passivity. In *Twenty Thousand Leagues*, first published in 1888, Verne predicted the swordfish-submarine conflict: "I saw there," says Captain Arronax, "some swordfish ten feet long, those prophetic heralds of the hurricane whose formidable sword would now and then strike the glass of the saloon." Underwater reality sometimes follows under-

Figure 4.5. USS *Swordfish*, commissioned in 1939, was the first submarine to sink a Japanese ship during World War Two. After 13 patrols in the Pacific, *SS-139* was sunk by the Japanese off Corregidor in January 1945. Richard Ellis collection.

water fantasy, for in July 1967, at a depth of 2,000 feet off Savannah, Georgia, an eight-foot swordfish charged the submersible *Alvin* and impaled itself in a joint between the upper and lower parts of the external hull. As the sub was brought to the surface, the fish was thrashing desperately, and as soon as they got a rope around its tail, the sword broke off. It took two hours to extract the sword, after which the fish was cooked and eaten. (The *New York Times* headline for January 14, 1967, was "Swordfish Duels Two-Man Research Submarine.")

Two years later, another swordfish attacked another submersible. On July 16, 1969 (the same day as the launching of *Apollo 11*, which put Neil Armstrong and Buzz Aldrin on the moon four days later), the mesoscaph (*meso* = middle; *scaphe* = boat) *Ben Franklin* was launched on a historic voyage, 30 days under water, following the Gulf Stream from Palm Beach, Florida, to a point south of Nova Scotia. Designed by Jacques Piccard (the pilot of the *Trieste* in 1960 on its record dive to 35,800 feet in the Marianas Trench), *Ben Franklin* was 48 feet long, with 29 viewing ports to permit maximum visibility for the crew of six. Somewhere off the coast of Florida, at a depth of 826 feet, three members of the crew watched as a six-foot-long broadbill swordfish attacked the submersible. In *The Sun beneath the Sea*, Piccard described the action: "It appears to parade up and down in front of the window, dashing to and fro, not knowing how to interpret our presence, swim-

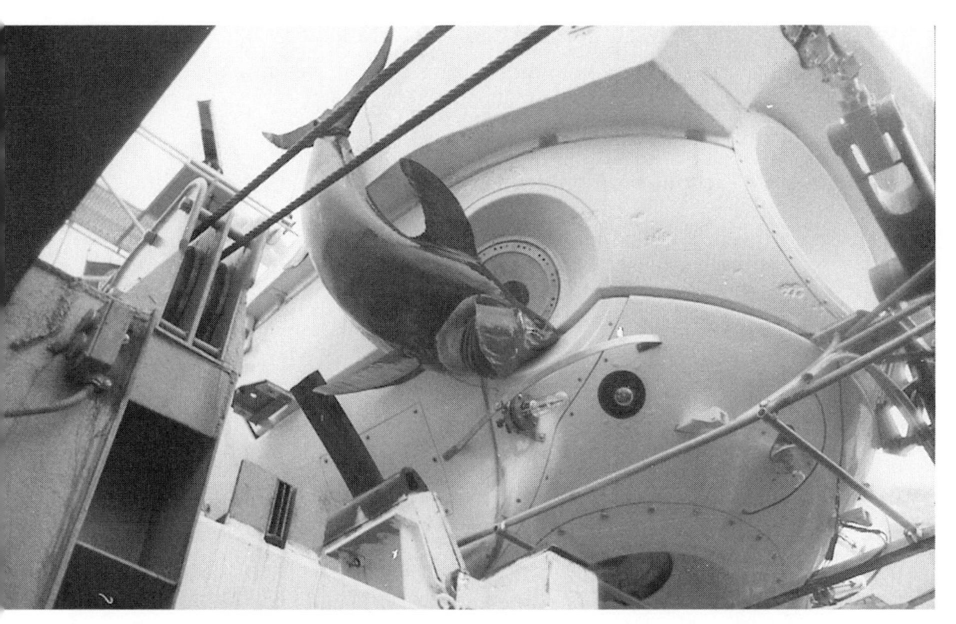

Figure 4.6. Off Savannah, Georgia, in 1967, at a depth of 2,000 feet, a swordfish charged the Woods Hole research submersible *Alvin* and died when its bill got stuck in the coaming. Photo courtesy of Woods Hole Oceanographic Institution.

ming a few meters and then returning, as though fascinated by our great Plexiglas eye. Suddenly it attacks, dashing straight forward and striking the hull of the mesoscaph with the point of the sword, aiming perhaps at the porthole, but hitting only the steel of the hull." Piccard asks: "Why do these swordfish attack submarines? Are they fascinated, hypnotized by the portholes? If they mistake submarines for monsters, which perhaps they actually are, what courage these fishes show in attempting to impale an adversary so much bigger than themselves."

As long as there are submersibles invading the swordfish's domain, it seems, the swordfish will resent them and respond accordingly. In August 2004, at 1,760 feet in the Gulf of Mexico, scientists aboard the *Johnson Sea-Link* (JSL) saw a fast-moving swordfish pass in front of their viewing port, reverse direction, and then hit the submersible with his (her?) sword. On the Harbor Branch website we can read,

> During their combined 8000+ dives the *Johnson Sea-Link* submersibles have come up against all manner of ocean creatures. Sharks,

other game fish, squid—they all take things calmly, either swimming on by or perhaps giving a little nudge. But not the swordfish. These things come in like they're convinced their little knife noses will be sufficient to take out a 23-foot, 14-ton submersible. They always lose, though one did once cut a hole in a hydraulic line. The real explanation for the behavior probably has something to do with the lights attracting then confusing the fish, but it's difficult to say.

I asked Don Liberatore, the *Sea-Link*'s pilot, about the frequency of the swordfish attacks. In an e-mail dated March 23, 2005, he answered:

> I think all of the active sub pilots have had an experience or two. We can think of 3–4 actual hits to the sub and 3–4 close encounters with that same aggressive or agitated behavior. That's not many compared to the 8000 plus dives that have been done by the JSLs. It seems like all of the attacks have been in the 1500–2200 ft depth range. All happen very quickly and so far with minor damage to the sub. The fish have broken off parts of their bills that have been wedged into the external equipment. It also seems that it's always in the front of the sub where most of the light is directed. I don't know how familiar you are with the JSLs but the front personnel compartment is an acrylic sphere that holds the pilot and a scientist. It could be that the animals are seeing a reflection of themselves in the sphere if the lighting is right and head towards that image. Or perhaps it's an escape behavior by heading towards the light. We have seen them at times motionless in the water then as we get closer zigzag very quickly back and forth and then shoot straight for us.

There is a video on YouTube (search "swordfish in oil rig"), of a large swordfish trapped in the underwater pipes and ladders of the blowout preventer of an oil rig in 2008. Fish can't back up, so this poor creature writhes and struggles with its sword stuck through the maze of pipes. From a remotely operated vehicle, technicians used a grabber to try to pull the fish out by the base of its tail, and they succeeded twice, but both times the fish darted right back into the rig. As you watch this incredible incident, you can't help but think that the fish was trying to attack the rig, especially after it has been freed and charges again. As

Figure 4.7. The original owner of this sword lost it—and its life—when it stabbed an oil pipeline off Angola in 2008. Richard Ellis collection.

the fish swims away, you can see that it is swimming erratically, and you can assume that its (self-induced) entrapment must have resulted in oxygen deprivation, but as far as we can tell, the fish lived to attack something else.

How does a swordfish decide what to attack? A careful look at the subjects of swordfish attacks—with the possible exception of those on whales—will reveal a pattern, or perhaps more than one. First of all, let us note that there are numerous records of unprovoked swordfish attacks on boats; where fishermen or recreational sailors looked down to find that their hull had been pierced by a passing swordfish (or marlin). Indeed, most of the "attacks" in Gudger's 1940 study are documented by the discovery of a broken-off sword in the hull or timbers of a vessel when it was careened. (Sometimes, of course, if the sword had been driven right through the planking, its point, and possibly part of the shaft, would then be prominently visible.) There is a certain similarity between those objects pierced that do *not* resemble the swordfish's prey, which usually consists of fishes or squid two feet long or smaller. Ships, whales, bales of rubber, or submarines in no way fit this description of prey, but they are all large objects, substantially larger than the swordfish. Though whales or bales of rubber could not be perceived as threatening to a swordfish, a fishing boat might be. This, however, presupposes an understanding of the function of a boat by the fish, which might be a bit of a stretch for the intellectual capacities of *Xiphias*. But Ralph Bandini thinks the swordfish is actually smarter than other fishes:

> He is the one fish that I have had anything to do with that really seems to think. Time and again, after hours of fighting, I have had them swim deliberately toward the boat, bring up about thirty or forty feet off, and lie there in the swells watching us, seeming to study us. And never once when that happens has the fish failed to change his tactics. Time and again when I have been circling one, I have had him keep turning away from me, keeping me on the outside of a circle, never giving me a chance to head him. Don't tell me they don't think, and figure things out for themselves!

Gudger, like almost everyone else, is at a loss to explain why swordfish attack vessels, but he offers some tentative suggestions. For attacks on vessels on the high seas, he suggests that prey fish such as albacore might "have the habit of sheltering in schools beside or under slow-moving vessels, and thus bring the vessel in line with the rush of a hungry swordfish. . . . This explanation is so simple and so sure that there is no necessity to say more of daylight attacks on slow-moving vessels making a passage or even standing still." He assumes that the swordfish was intent on *spearing* the prey fish and made a headlong charge, sword extended, rather like a jouster with his lance. But spearing would seem to be a particularly ineffectual method of feeding, and therefore Gudger's explanation doesn't seem quite so "sure and simple." As for dories, Gudger concludes,

> That a swordfish, struck in or near the head (in the spinal column or brain) would lose its sense of equilibrium is readily understandable. That such a fish held by a line would gyrate around in circles with that line as a radius seems sure. That such a stricken fish when hauled in toward the dory, in order to ease the pain increased by the pull on the line on the harpoon embedded in the flesh, would follow the warp straight toward the boat is understandable. That this fish when close enough to perceive the boat ahead of it would associate it with the cause of its pain and the loss of freedom, seems not improbable. If these are sound conclusions as based on the facts, then we may go further and expect the attack to follow, and it would then look like a deliberate one.

I'm not so sure that Gudger's conclusions are sound. I don't know that a struck fish loses its equilibrium; or that it would head toward the dory to "ease the pain"; and, to me, it is most improbable that a fish would associate a dory with "the loss of freedom." I think that swordfish attacks on boats look deliberate because they *are* deliberate, but I don't know why they behave that way. Neither does Bandini. In the his 1933 article on swordfishing, he wrote that the swordfish "is an individualist, if ever there was one, who never does the same thing twice running, who may sound hundreds of feet and sulk there all day and all night, or who may charge your boat time after time, sword out, great jaws wagging, the picture of insensate fury, or who may leap in great awkward lunges, throwing white water 10 feet in the air as he smashes back, or who may deliberately swim right up to your boat and calmly look you over as though he were determining by observation just what tactics he will pursue."

In a highly publicized account—it was the cover story of the September 2005 issue of *SaltWater Sportsman*—a black marlin "attacked" a fishing boat off Panama. Hooked by 18-year-old Stephen Schultz, the 500-pound fish "made an impressive aerial display just 30 feet off the stern," submerged briefly, and then erupted from the water again and began tail-walking right toward the boat. It didn't make it over the transom, but it did make contact with the fisherman, knocking him from his chair and somehow breaking four bones in his face and lacerating his throat. The article, written by David Dibenedetto, editor of the magazine, suggests that the hooked fish attacked its tormentor, but he also directs the reader to the magazine's website (saltwater sportsman.com) where a video of the "attack" can be seen. I've seen it, and it looks not so much like an attack to me but, rather, a fish trying to escape the pain of a hook in its mouth, tail-walking in a panic, and accidentally crashing into the boat.

Now the score is marlins 2, fishermen 0. In August 2006, as fisherman Leslie Spanswick was trying to reel in a marlin (estimated at 14 feet in length) during a fishing tournament out of Bermuda, the fish leapt out of the water and speared mate Ian Card in the chest. In an article in *SaltWater Sportsman*, Card recollected, "I remember standing in the cockpit. I remember seeing the fish in the air, just for a moment, as it sailed over the right corner of the stern. I remember moving a step to the side just before the bill speared me in the chest." He was knocked overboard and, still attached to the fish, dragged underwater. He pulled himself free of the marlin's bayonet and was pulled back aboard the *Challenger*. The bill had passed through his chest, and the tip came out of his back. A towel was stuffed into his "wound as big as your fist," and he was rushed to Bermuda's King Edward VIII Hospital for emergency surgery. Doctors said that if the wound was half an inch higher, it would have severed an artery and killed him (*Daily Mail* [London], August 4, 2006). If we count the swordfish attack on Mark Ferrari—where the wound was remarkably similar—the score is actually billfish 3, humans 0. But if you count *fishing*, the score is a rather more lopsided: humans 11,675,398 (or thereabouts), billfish 4.

It is possible to visualize a swordfish charging into a tightly packed school of fish, slashing as it goes with its sharp sword, but the rounded bill of the marlins presents more of a problem. Does the marlin wield its sword like a club, clobbering fishes senseless before gobbling them up? All the Istiophorids are powerful swimmers—and they have teeth—so they can also feed by chasing down their prey. They are known to prey on fast-swimming, relatively large fishes such as dolphins (*Coryphaena*) and small tuna, so they must be able to catch and capture them in the open sea. Pelagic squids, also fast-swimming creatures, are capable of leaping out of the ocean to escape, but they too have been found in the stomachs of captured marlins. Nakamura reports that a fisheries biologist working the Sea of Cortez off Baja California saw a large blue marlin (*M. mazara*) approach a school of Humboldt squids (*Dosidicus gigas*) "that was gathering under the night light of a squid-fishing boat; it approached the school at almost full speed with its fins completely held back in the grooves, then suddenly hit the squids with its bill, subsequently nudging the stunned prey and eating it head first." Nakamura further noted that the Indo-Pacific marlin "has also been observed to swallow big tunas like skipjack (*Katsuwonus pelamis*), yellowfin tuna (*Thunnus albacares*), and bigeye tuna (*Thunnus obesus*) head first, and the fishes found in the stomachs of *M. mazara* often showed deep slashes on their bodies, presumably caused by the bill of the marlin." A picture of the great marlin as a superpredator is beginning to emerge: this is a fish that will eat anything it can catch, and it can catch just about anything that swims.

The newest record belongs to fisherman John Barfield, fishing out of Lighthouse Point, Florida (south of Boca Raton on the Atlantic coast), in May 2009. After hooking a large swordfish (later weighed at 355 pounds), Barfield was trying to bring it aboard the boat, when the fish flipped itself out of the water and crashed into him. His collarbone was broken, he received a three-inch gash in his forehead, and the hook, originally in the mouth of the swordfish, ended up deep in Barfield's thigh. Naturally, the local newspaper coverage headlined this event "Fisherman Attacked by Swordfish." Never mind that the fisherman actually attacked the fish in order to kill it, and the fish was doing nothing more aggressive than fighting for its life.

Because a swordfish regularly uses its sword, and because a charging swordfish arrives *sword first*, attacks should not surprise us all that much. As there would be no benefit to the fish in spearing the ocean floor, it is reasonable to conclude that the harpooned fish that drove its sword into the mud had been driven mad by the pain and was diving to escape. The "attack" on Mark Ferrari occurred while the swordfish itself was under attack and ready to strike at anything—especially something as awkward and defenseless as a human diver. Attacks by swordfish that are not hooked or harpooned are a little more difficult to explain, but, as Jordan and Evermann wrote, "driving their swords through planking [may be] 'temporary insanity,' or more likely, while pursuing dolphins or other fish." I must admit that I can come up with no explanation that would explain a swordfish attacking an 80-foot-long whale or a bale of rubber, but the attacks on the submersibles *Alvin* and *Ben Franklin* might be understood if we recognize that swordfish at depth are stimulated by lights.

Swordfish hunt bioluminescent lantern fishes at depth, slashing at the blue lights in an effort to maim or incapacitate as many of the little fishes as possible, and we know that swordfish feed on *Dosidicus*, a large squid species known to flash brightly from red to white. Is it too much of a stretch to hypothesize that the lighted portholes of the submersibles would flip a switch in the swordfish's (overheated) brain and provoke the "attack" response? Neither whales nor submarines resemble anything that the swordfish might take for a predator, and because the concept of a prey item preemptively attacking a potential predator is unknown in the animal kingdom, attacks on these objects remain a complete mystery. (People, of course, are the swordfish's ultimate predators, but the fish doesn't find that out until it's too late.)

On the African plains, lions are always predators and zebras are always prey. The same is true for tigers and deer in India, or jaguars and capybaras in South America. At no time does the zebra turn the tables and attack the lion, or the deer the tiger, but in the ocean there are no such clear predator/prey distinctions. As the fishermen says in Shakespeare's *Pericles*, "the great ones eat up the little ones" and so ad infinitum, but it is only the act of eating the little fish that defines the bigger fish as a predator. Thus a larval swordfish, a couple of inches long, might be eaten by mackerel or bonito, among the recognized prey species of the adult swordfish. (The mackerel sharks—mako, great white, porbeagle—begin their lives as fully formed predators. A newborn great white, for example, is about four feet long, already larger and better equipped than most other shark species as adults.) As we've seen, a school of false killer whales will attack an adult swordfish, and Jordan and Evermann wrote that "sperm and killer whales and the larger sharks alone could menace them." (The operative word there, is "could." There are no records of sperm or killer whales feeding on swordfish.) They wrote that there is one account of "a 120-pound swordfish nearly intact with sword still attached, found in the stomach of a 730-pound mako taken near Bimini, Bahamas." Mako sharks can get to be as big as the swordfish—the IGFA record mako weighed 1,221 pounds—and they are probably as fast. At the front end, where the swordfish has a broadsword, the mako has a mouthful of scalpel-sharp teeth. Both species are solid muscle.

On July 28, 1977, David Lester and his wife Kim were fishing for swordfish about 20 miles south of Montauk, Long Island, and although they spotted two swordfish, they were unsuccessful in harpooning either. They spotted a huge shark and after harpooning it, they were towed all over the ocean for three hours before the shark tired and they got a tail rope on it. The shark, a female mako, was 11 feet four inches long and weighed 1,250 pounds. (This was 39 pounds heavier than the world's record mako, but obviously ineligible for an IGFA record because the shark was harpooned, not caught on rod and reel.) Packed in ice, the shark was transported to the NMFS Laboratory in Narragansett, Rhode Island, to be examined by shark specialists Jack Casey, Wes Pratt, and Chuck Stillwell. I was there when the shark's distended stomach was cut open, and I was as surprised as everyone else to see 79 pounds of fresh swordfish in volleyball-sized bites. One chunk contained part of the skull and the base of the sword; the weapon of the swordfish had been bitten completely in two. Chuck Stillwell (one of the authors of the 1985 study of the food and feeding ecology of the swordfish in the western North Atlantic) was quoted in the *Long Island Traveler-Watchman* as suggesting that "the mako struck first in the head region, snapping the bill which would have been no easy feat, for the swordfish was probably over 350 pounds. . . . The shark next took bites from the [dorsal] fin region, crushing the backbone as she ate."

The mako is one of the very few marine creatures equipped or inclined to prey on an adult broadbill, and, according to Frank Mundus and William Wisner, "makos and swordfish are mortal foes." It would appear, however, that enmity is in the eye of the beholder. Fishes that attack other fishes do so only out of a desire for food, not because the prey is an "enemy." (Most shark attacks on people are probably inspired also by hunger; and *Jaws* notwithstanding, great white sharks do not attack people out of malice.) In his book *Sportfishing for Sharks* (with Wisner), Captain Frank Mundus (who is said to have been the model for Captain Quint in *Jaws*) relates the story of an angler "battling a superb swordfish when a large shark—probably a mako in this

Figure 4.8. Aside from fishermen, the adult swordfish's only enemy is the mako shark. Painting by Richard Ellis.

case—barreled in out of left field and amputated the better part of the broadbill's rear section." In *Shark, Shark* (with H. Z. Mazet), Captain William Young tells of a case in which the roles were reversed, and a dead shark was found on the beach with "18 inches of a Broad-Billed Swordfish sword in the vital organs of the shark."

When the swordfish "attacked" Mark Ferrari, it might very well have been reacting to the presence of the false killer whales. Although the swordfish probably didn't know it, false killer whales are in the habit of preying on swordfish—as long as the swordfish are dangling from longline hooks. Ramos-Cartelle and Mejuto studied the depredation of false killer whales on swordfish by the Spanish longline fleet in the Atlantic, Indian, and Pacific Oceans and wrote that "depredation may affect a number of swordfish equivalent to 50% or more of the catch held on board, and may even damage the catch in a proportion that is several times greater than the number of swordfish retained on board."

Of course, long-liners catch more than swordfish, but the authors estimated that about 2 percent of all worldwide swordfish catches were depredated by *Pseudorcas*.[1]

Xiphias the gladiator has no scales, no teeth, no pelvic fins, and no lateral line. It does have a brain warmer, an eye warmer, a bayonet, and a reputation for pugnacity unmatched in any other fish. This reputation, in fact, may be unmatched in any other animal. Some sharks occasionally "attack" some people and, less frequently, eat them, but while this behavior may be characterized in many ways—hunger, territoriality, mistaken identity—it is probably not aggression for aggression's sake. So too with lions, tigers, or crocodiles, often described as "man-eaters" but trying to catch something to eat is not aggression, no matter how violent and bloody. There are many animals whose bite, sting, or spines are venomous, and while people can die from the bites of snakes or spiders, these creatures are usually acting in self-defense or possibly in defense of their offspring or nest. No other animal seems willing to go out of its way—and often to endanger its own life—to attack. As we have seen, swordfish have been known to attack whales, sharks, fishing boats, submarines, bales of rubber, and people—not to mention the variety of fish and squid species that they "attack" in order to eat them or the pieces.

For most of recorded history, when men went fishing, it was to catch something to eat. Around the turn of the twentieth century, however, the notion of fishing for *sport* arose. There is, of course, no particular challenge in capturing a carp or a codfish, and while some will argue that the capture of a salmon requires great skill and finesse—not to mention the art required to create a "fly" that will fool the fish—fish the size of most carp, codfish, or salmon do not present that much of a *physical* challenge. But if the fish weighs 1,000 pounds and must be located and caught far offshore, often with expensive tackle and

1. Along with the photographs of "depredated" swordfish, tuna, and sharks in this paper, there is a photograph of a dead false killer whale that is captioned, "One of the false killer whale specimens that suffered incidental mortality in the Atlantic Ocean." Occasionally, the false killers get hooked by the flippers or tangled in the lines and drowned. There is no free lunch, even when it's hung on a line in front of you.

from an expensive fishing vessel staffed with specialized crew members, this new sort of fishing becomes something very different indeed from subsistence fishing. It is "big-game fishing," a sport (and a challenge) originally reserved for men (and an occasional helpmeet) who have enough time, money, and passion to "hunt" giant fish in exotic and far distant seas.

5 Sport-Fishing for Swords

Lassoing mountain lions, hunting the grizzly bear, and stalking the fierce tropical jaguar, former pastimes of ours, are hardly comparable to the pursuit of Xiphias gladius. It takes more time, patience, endurance, study, skill, nerve, and strength, not to mention money, of any game known to me through experience or reading. . . . I have caught two in nine years, and have seen more and hooked more than any angler since Boschen. I have made a special pursuit and study of Xiphias gladius, and have had a boat built—the Gladiator—just for that purpose. And I have only begun to appreciate the strangeness, intelligence, speed, strength, and endurance of this king of the sea.—Zane Grey, *Tales of Fishes*

By the latter part of the nineteenth century, the broadbill swordfish was recognized as one of the world's premier game fishes and had also gained a reputation for unmatched pugnacity. G. Brown Goode, assistant secretary of the Smithsonian and a prodigious author and editor of American fisheries literature, wrote this about "the perils and romance of swordfishing" in 1887:

> The pursuit of the swordfish is much more exciting than ordinary fishing, for it resembles the hunting of large animals upon the land, and partakes more of the nature of the chase. There is no slow and careful baiting and patient waiting, and no disappointment caused by the accidental capture of worthless "bait-stealers." The game is seen and followed, and outwitted by wary tactics, and killed by strength of arm and skill. The swordfish is a powerful antagonist sometimes and

> sends his pursuers' vessel into harbor leaking, and almost sinking, from injuries inflicted by a wounded swordfish. I have known a vessel to be stuck by a wounded swordfish as many as twenty times in one season. There is even the spice of personal danger to give savor to the chase, for the men are occasionally wounded by the infuriated fish.

In those days, swordfish were harpooned. It is believed that the first swordfish ever taken on rod and reel was caught in 1913 by a fisherman named William Boschen, off the California coast at Catalina. Boschen, who was identified by Zane Grey in 1919 as "probably the greatest heavy-tackle fisherman living," was a member of the Tuna Club, formed in 1898 by a group of fishermen who fished out of Avalon. A guide called George Farnsworth had developed a system for getting the bait away from the boat, which consisted of an ordinary kite flown downwind, with the bait (and hook) suspended from the kite string. This meant that the bait wasn't being bounced through the wake of the boat which deterred curious fish from biting and proved to be extremely successful for the man who tried it. (The current solution to this problem is to use "outriggers," long, flexible poles that stand out far from the boat.) Boschen's first broadbill weighed 355 pounds; several years later, he caught another one that weighed 463 pounds, which was the world's record at that time. Boschen was obsessed with swordfish, but no more so than his fellow Tuna Club member, Zane Grey.

Born in Zanesville, Ohio, in 1875, Zane Grey was originally christened "Pearl," but he felt that was a sissy's name, so he changed it to reflect his birthplace. Like his father, he was trained as a dentist but, realizing that his future lay elsewhere, he turned to writing. In the early decades of the twentieth century, he wrote enormously popular western novels, many of which were made into equally popular movies. Grey became a passionate big-game fisherman, spending much of his not inconsiderable royalties (his books have sold 13 million copies) on fishing trips, fishing boats, fishing camps, and fishing gear. For a while, the scientific name of the Atlantic sailfish was *Istiophorus greyi*, but this name is no longer valid, and his claim to immortality now depends on his books. In addition to books like *Riders of the Purple Sage*, *Code of the West*, and *The Westerners*, he wrote about his angling exploits in *Tales of Fishing Virgin Seas* (1925), *Tales of Swordfish and Tuna* (1927), *Tales of Tahitian Waters* (1931), and *An American Angler in Australia* (1937).

His skill as a writer of fishing stories is unequalled; he is probably the best big-game fishing writer who ever lived. Hemingway was a writer who fished, but Zane Grey was a fisherman who wrote—and wrote superbly. In a review of a 2005 biography of Grey, Jonathan Miles said that "Grey's fishing essays are largely devoid of the pinched overexertion of his fiction, and are actually quite excellent." In these fishing essays, he was given to unreconstructed hyperbole and the sort of exaggeration that can be said to define those who would write about fishing, but no one has ever approached his ability to express the excitement, the glory, and the heartbreaking failures of big-game fishing.

Grey's first published "outdoor" book was *Tales of Fishes*, published in 1919, in which he chronicled his adventures in Mexico, California, and Florida from 1914 to 1919. This book contains chapters on fishing for tarpon in Mexico; for wahoo, barracuda, and sailfish in the Gulf Stream; and for bonefish in the Florida Keys; but its main emphasis is Grey's passion—Grey's *obsession*—with the billfishes.[1] In 1915, the fish we now call the marlin was known as swordfish, roundbill swordfish, or marlin swordfish, and Grey devotes several chapters to the pursuit of marlins in California waters. But in *Tales of Fishes*, he acknowledges the broadbill as the greatest challenge of all:

> Three summers in Catalina waters I had tried persistently to capture my first broadbill swordfish; and so great were the chances against

1. The book also contains a chapter called "Swordfish," written by the aforementioned George Brown Goode, in which we learn many of the more arcane details of the natural history of *Xiphias gladius*, including its range and size, the history of the fishery, and the nature of the harpoons. Goode's chapter is identified as having come from the "New York Bureau of Fisheries," but much of it appears verbatim in Goode's chapter on swordfish fishery in his 1887 *The Fisheries and Fishing Industry of the United States*. Tacked on to Goode's chapter in *Tales of Fishes* are some facts about the sailfish, the spearfish, and the cutlassfish. (The cutlassfish, *Trichiurus lepturus*, is not a billfish at all, but a four-foot-long, elongated, flattened ribbon of a fish, with a head like that of a barracuda.)

Figure 5.1. Zane Grey poses proudly with his 582-pound swordfish, caught off Avalon, California, in 1926. Photo courtesy of Ed Pritchard—AntiqueFishingReels.com.

me that I tried really without hope. It was fisherman's pride, I imagined, rather than hope that drove me. At least I had a remarkably keen appreciation of the defeats in store for any man who aspired to experience that marvel of the sea—*Xiphias gladius*, the broadbill swordsman. . . .

The broadbill swordfish is a different proposition. He is larger, fiercer, and tireless. He will charge the boat, and nothing but the churning propeller will keep him from ramming the boat. There were eight broadbill swordfish hooked at Avalon during the summer, and not one brought to gaff. This is an old story. Only two have been caught to date. They are so powerful, so resistless, so desperate, and so cunning that it seems impossible to catch them. They will cut bait after bait off your hook as clean as if it had been done with a knife. For that matter, their broad bill is a straight, long, powerful two-edged sword. And the fish perfectly understands its use.

On his first approach to New Zealand in 1925—remember, big-game fishermen in those days had to sail their fishing boats to where they wanted to fish—Zane Grey was taunted by the fish that would subsequently become his totem. Early in *Angler's Eldorado* (the account of his New Zealand adventures), he wrote:

> While watching an albatross, I was tremendously thrilled by the sight of an amazingly large broadbill swordfish. He was not over three hundred yards from the ship. His sickle fins stood up strikingly high, with the old rakish saber shape so wonderful to the sea angler. Tail and dorsal fins were fully ten feet apart. He was a monster. I yelled in my enthusiasm, and then ran for captain Mitchell. But on my return I could not locate him. The fish had sounded or gone out of sight. This was about fifteen miles offshore; and it was an event of importance. Swordfish do not travel alone.

Romer Grey—known as R. C.—was evidently drawn into big-game fishing by his older brother. As Zane Grey wrote in his introduction to his brother's book, *Adventures of a Deep-Sea Angler*, "Now R.C. had not a single drop of salt blood in his veins. The idea of fishing in the sea was obnoxious, ridiculous, and impossible for him." Regardless, Zane was going to make a fisherman out of Romer, and year after year, he dragged him along. At one time, Romer hooked a swordfish off Avalon and fought "him" for more than five hours before he tired and had to hand the rod to his brother. Zane Grey fought the fish, and when he tired, he handed the rod to the skipper, Captain Dan. In the darkness of evening, after six more hours of trying to reel in the giant fish, Romer "heard a delicate spattering from everywhere and I saw against the sky, in the gleam of the deck light, a silver flash. Flying fish! 'He's chasing flying fish!" And so he was. After fighting three fisherman and a hook in his mouth for more than 11 hours, the "foxy unbeatable old broadbill was feeding on flying fish, and having a grand time." In recognition of such a show of power—and massive indifference—Captain Dan dropped the rod tip and purposely let the line break, setting loose the fish. It is almost impossible to imagine a fish so strong that it casually feeds after 11 hours of being dragged all over the ocean by three men in a boat—or even dragging the boat for some of that time.[2] It is a testimony to Zane Grey's respect for the broadbill that he let that one go.

It would not be an exaggeration to say that Zane Grey's fishing tales are dominated by his quest for the broadbill—indeed, it would not be much of an exaggeration to say that a good part of his *life* was dominated by his quest for the broadbill. There are virtually no limits to his celebration of this fish; paeans to *Xiphias* fill his fishing books, and could easily fill this one. In *Tales of Swordfish and Tuna*, he wrote:

> Old *Xiphias gladius* is the noblest warrior of all the sea fishes. He is familiar to all sailors. He roams the Seven Seas. He was written about by Aristotle 2,300 years ago. In the annals of sea disasters there are records of his sinking ships. . . . Tales of his attacks on harpooners' boats in the Atlantic are common. In these waters, where he is hunted for the market, he has often killed his pursuers. In the Pacific, off the Channel Islands, he has not killed any angler or boatman yet, but it is a safe wager he will do so some day. Therefore, despite the wonderful nature of the sport, it is not remarkable that so few anglers have risked it.

2. The endurance record—positive for the fish, negative for the fisherman—is held by a fisherman named D. B. Heatley, who in 1968, fought a swordfish for 32 hours and five minutes before the fish escaped.

Of course the intrepid Grey was more than a little impressed by a fish that fights back—he welcomed the challenge. In 1919, off Catalina Island, he hooked a swordfish:

> First he made a long run, splashing over the swells. We had to put on full power to keep up with him, and at that he took off a good deal of line. When he slowed up he began to fight the leader. He would stick his five-foot sword out of the water and bang the leader. Then he lifted his enormous head high and wagged it from side to side, so that his sword described a circle, smacking the water on his left and then on his right. Wonderful and frightful that sweep of sword! It would have cut a man in two or have pierced the planking of a boat. Evidently his efforts and failure to free himself roused him to a fury. His huge tail thumped out of great white boils; when he turned sideways he made a wave like that behind a ferryboat; when he darted here and there he was as swift as a flash and he left a raised bulge, a white wake at the surface. Suddenly he electrified us by leaping. . . . This one came out in a tremendous white splash, and when he went down with a loud crash, we all saw where the foam was red with blood.

After 11½ hours, the line went slack, and Grey lost his giant swordfish. He expressed his admiration for this most worthy of opponents by continuing to fish for them and, later, proudly brought in a 418-pounder. It weighed 45 pounds less than Boschen's world record, and Grey never forgot it. In *Angler's Eldorado*, originally published in 1926, Zane Grey recounts his ongoing hunt for the mighty broadbill. As of 1926, he had managed to acquire the world's record, which was a 578-pounder caught in the North Atlantic. When he realized that the hooked fish was a *broadbill* (his italics), he wrote, "The second leap was enough to dazzle any boatman, let alone those who had never seen a broadbill. It was a forward jump, quite high and long, allowing us time to see his bronze bulk, his wide, black tail, his huge, shiny head, and waving sword." (Grey referred to every big fish as "he," even though all large swordfishes are females.) After almost an hour, Grey brought the 578-pound fish to the boat. "Loading the swordfish," he wrote, "we ran in to exhibit him to seven or eight boats fishing there. I shall not soon forget the expression of those anglers. Such a marvelous and amazing fish as the broadbill had never been imagined by them." That same year, Grey bested his world's record by four pounds, reeling in a 582-pounder of Catalina, California. (Nobody, least of all Zane Grey, could have imagined that Lou Marron, fishing off Chile in 1953, would haul in the current world's record swordfish—at 1,182 pounds, more than twice as heavy as Grey's record.)

In his *Tales of Fishing Virgin Seas,* Grey ruminates on the giant broadbills they would never see:

> Broadbill swordfish frequented the Gulf [of Mexico], but never in great numbers. [Captain Billy Clover] harpooned one that was over twenty feet long and red in color. While he was towing it in the wind came up hard, making the sea so rough that the swordfish was torn away. This does not seem to me to be an exaggeration. Old *Xiphias gladius* grows to enormous size and roams all the Seven Seas. Around the Marquesan Islands swordfish attain huge dimensions, according to some travelers. The natives there tell of swordfish with sword[s] seven feet long. They attack and sink canoes, and kill fishermen. The Cubans also tell of a very large and dangerous swordfish.

Other big-game fishermen have been bitten by the broadbill bug and have traveled halfway around the world in pursuit of the glorious gladiator of the sea. In 1940, an expedition from the American Museum of Natural History in New York set out to capture some of the "fighting giants of the Humboldt," as they were termed in the title of the *National Geographic* article by David Douglas Duncan. Because of the upwelling of nutrients, the Humboldt Current off the west coast of northern South America is inhabited by hordes of baitfish, which attract swordfish and marlin, the largest of all pelagic predatory fishes. The AMNH expedition consisted of its sponsors, Michael and Helen Lerner, and Francesca LaMonte of the museum's ichthyology department. Fishing off Tocapilla, Chile, they caught several broadbills, ranging from 570 to 630 pounds and set the stage for fishing in the area now recognized as the best place in the world for big swordfish.

It is also one of the best places in the world for jumbo squid. Early in the commercial *Dosidicus* fishery, most of the catch was used for bait, but in recent years, the production of frozen *Dosidicus* products,

Figure 5.2. Louisburg, Nova Scotia, August 1936. Michael Lerner poses with two swordfish he caught in one day; a 535-pounder and a 601-pounder. International Game Fish Association.

Figure 5.3. Michael Lerner, Tocopilla, Chile, 1936. International Game Fish Association.

known commercially as *daruma*, has increased greatly. There was no fishery in Chilean waters for *jibia gigante* from 1971 to 1991, but the increased demand for squid filets has now resurrected the fishery, and by 1994 landings had reached 190,000 tons and produced $18 million for the Chileans. The squid are also harvested in Peruvian and Mexican waters and sold primarily in China and Japan. The Gulf of California *Dosidicus* fishery is now conducted by three fishing fleets that move around the Gulf in pursuit of the migrating squid. It has been estimated that the total population of *Dosidicus* in the Gulf of California is around 171,000 tons, but because squid like *Dosidicus* reproduce quickly, a depleted population can replenish itself quickly and effectively. Like most squid species, *Dosidicus* grows fast and dies young. In a 2004 study of age, growth, and maturation of the jumbo squid, Markaida et al. noted that *Dosidicus* shows the fastest growth rate of any squid species, but they don't live very long—the larger specimens may get to be two years old, but most complete their life cycle in less than a year.

After catching the jumbo squid, the AMNH expedition quickly shifted its focus—as did Duncan's 1940 article—from giant fish fishing to giant squid fishing. When the anglers learned from their Peruvian guides that the waters they were plying were also home to monster squid ("Those things can't be taken on rod and reel!" said Juan), they outfitted themselves in pillowcases with eye holes to protect them from the powerful jets of ink, and, according to Duncan, "began one of the strangest adventures in the history of scientific collecting and deep-sea angling." In their bizarre Ku Klux Klan–like outfits, the fishers hauled up the ten-foot long, 100-pound squid and became the first Americans to watch these animals in action. As the cephalopods darted through the dark waters, they created swaths of phosphorescence that Duncan described as "incredibly fast-moving streaks of fire [that] closed in from all sides." Francesca LaMonte—AMNH ichthyologist and author of several popular books on big-game fishes—examined the stomachs of the swordfish that were caught and found that one's stomach was "so full of squid that she marveled at its having been able to swim, let alone

Figure 5.4. Off the coast of Peru in 1940, big-game fisherman Michael Lerner (*right*) went looking for swordfish but, instead, he started reeling in aggressive, eight-foot-long Humboldt squids. International Game Fish Association.

show interest in the bait." It is not clear from Duncan's description if all the squid in the swordfish's stomach were *Dosidicus*, but they probably were, and as the only squid species mentioned, one would like to know their condition. Were they swallowed whole, or sliced and diced? The manner in which swordfish feed is only one of the mysteries surrounding this great and enigmatic fish.

Ralph Bandini (1884–1964) served on California's Fish and Game Commission and on the board of directors of the Tuna Club of Catalina Island. As with so many deep-water fishermen, he revered the swordfish above all other species. In his 1939 book, he tries to explain why:

> In my opinion, of the Big Three—tuna, marlin and broadbill—the latter is by far the hardest to hook, to handle, or to take. His vitality is beyond belief. Let me give you an example. The late W. C. Boschen, who took the first broadbill on rod and reel and who was the father of the sport, if you want to call it sport, took an average sized fish after about a three or four hour fight. When the fish was gaffed they hauled him up upon the stern of the boat and lashed him down. The run back to Avalon consumed about three and a half hours. There was another half hour used in hoisting the fish from the boat to the weighing standards. Just as they were about to weigh him he came to life. He started threshing around, broke the rope by which he was hung, smashed over one of the standards, which weren't toys, ran everybody off the end of the wharf, knocked a piece out of the rail, and plunked back into the water! And all that, mind you, after he had been out of the water for over *four hours*.[3]

As fishing writer Pat Smith wrote in 1973, "The word broadbill is a guaranteed showstopper in the posh big-game playpens of the western world. Whether you're sitting at the bar at Walker's Cay in the Bahamas, the Club de Pesca in Panama, or the Deep-Sea Club in Montauk,

3. For a writer about big-game fishing, Bandini was unusually truthful and reliable, but like Zane Grey, Hemingway, and many other swordfishermen, he believed that all the biggest swordfish were males. ("Just as they were about to weigh him he came to life. He started threshing around.") And indeed, no swordfish could possibly remain out of the water for *four hours* (Bandini's italics) and survive.

New York, the odds are that you're unique among your companions if you've caught a broadbill, and you're privileged if you've even seen one. For despite his vast numbers and global range, the broadbill is perhaps the most elusive game fish in the ocean." The broadbill swordfish is the holy grail of sport fishing. It is a testimony to it strength, size, and elusiveness that nearly every large specimen has been lovingly recorded, something that can be said of no other fish—except perhaps the very large marlins. Along with the marlins and bluefin tuna, broadbills can grow to enormous size, which makes them the worthiest of opponents in the sport of fishing. That the fish often wins the battle—for which read: escapes—makes them that much more exciting. (And with the swordfish, there is the not insignificant possibility that the fish will turn the tables and attack the fisherman.) As George Reiger wrote in *Profiles in Saltwater Angling*, "In fishing around, Lerner heard much talk about a single gamefish that was revered above all others. Kip Farrington made a virtue of the fact that he had fished for six seasons before catching his first. Even Ernest Hemingway could be shaken in his devotion to giant blue marlin when conversation turned to the legendary broadbill swordfish."[4] Most fishermen have never even seen the twin sickle fins of a sunning broadbill, let alone hooked one. So when a place was found where very large swordfish could be found in abundance, rich Americans packed up their rods and reels and headed south. The place was Tocopilla, in northern Chile, a couple of hundred miles south of the Peruvian border.

Fishing out of Tocopilla in 1934, an American named George Garey caught two broadbills in the same day, each of which weighed approximately 500 pounds. The same year, W. E. S. Tuker, an Englishman who ran the Anglo-Chilean Railroad, caught a swordfish that was 13 feet long and weighed 837 pounds. In 1939, Tuker invited Kip Farrington

4. Not so, according to Paul Hendrickson, who wrote *Hemingway's Boat* in 2011: "Once he'd found marlin, all bets were off; it was as if every other creature in the sea was just guppy sport. Yes, giant bluefin tuna and broadbill swordfish, and, to some extent, mako shark off Bimini in the mid-thirties would have their obsessions. But for the rest of his life, marlin reigned supreme, most especially the blue marlin: *Makaira nigricans*, a trophy that could go to fifteen feet, could go from fifty pounds to twelve hundred."

Figure 5.5. This 853-pounder was caught by Kip Farrington, Tocopilla, Chile, 1941. International Game Fish Association.

and his wife to fish there, and, as Reiger wrote, "the Farringtons—fishing separately—sighted no less than 73 broadbill swordfish, presented baits to 37 of them, had strikes from 26, hooked 19, and caught four." Tuker apologized to the Farringtons, as "he had never known anyone to have poorer luck while fishing off Chile." One day in 1940, fishing out of Montauk, Long Island, Farrington "foul hooked" a swordfish through the dorsal fin. The fish swam off, evidently so unperturbed by the hook through its fin that upon encountering a school of bonito it began to feed. "It was just like having a dog on a leash," wrote Farrington. After feeding, the 300-pound swordfish began chasing the boat, but after about half an hour, the fish gave up the chase, and at 4:30 in the morning, after fighting the fish for more than eight hours, Farrington brought it alongside the boat. "All of my friends gave me a

Figure 5.6. A swordfish trying not to become a trophy or a dinner. Note the disproportionate length of the sword. Photo: Richard Stanczyk, Bud n' Mary's Marina, Islamorada, Florida.

good kidding on the length of time I took, particularly Mike Lerner and Ben Crowninshield, but I would like to have seen them done much better, especially with a Montauk swordfish." Chisie Farrington did all right with her Montauk swordfish; in July 1940, she caught "the largest fish ever taken by a woman off Montauk," a 298-pound swordfish. The following year, Farrington went back to Chile, where he caught five swordfish, the largest weighing 853 pounds.

Like Zane Grey, Kip Farrington was an author and a fisherman, but Farrington's books were nonfiction and mostly about fishing. (He also wrote about railroads.) One of his later books was *Fishing with Hemingway and Glassell*, in which he recounted this exchange: Hemingway told him that the two biggest thrills in a man's life were his first bullfight and his first sexual intercourse, and Farrington responded with, "Of the two greatest thrills in a man's life, *first* was his first broadbill swordfish, and *then* his sexual intercourse. And the latter was a whole lot easier to get!" Farrington was a purist, who believed there was a right way and a wrong way to catch fish. When he learned that there were scoundrels who were harpooning swordfish for "sport," he was outraged. In his book *Fishing the Atlantic*, he wrote:

> I know wealthy men who harpoon fish and then bring it in and sell it. I know others who harpoon only for the "sport" of it, and I know others who harpoon a swordfish after it has refused to strike a bait the two or three times it is presented. What thrill they can get out of harpooning this great sporting fish whose habit it is to swim slowly along the surface of the ocean during certain hours of the day is beyond me. It just isn't sport. Commercial fishermen have been in the business for eighty years more, and I don't begrudge them any fish they wish to harpoon. But they are food merchants. Of course, the price is what makes a man harpoon swordfish.

Linda Greenlaw, who long-lined swordfish for a living, had a different take on harpooning. If she didn't have to catch them by the hundreds, she said, she would prefer to harpoon them. "The harpoon is primitive," she says, and "the harpoon fishermen [are] among the last of the true hunters. Virtually anyone can be taught to bait a hook, but few possess the skill, concentration, and coordination to iron a free-swimming fish."

Captain Greenlaw notwithstanding, the last of the macho fishermen was Ernest Miller Hemingway, born in Oak Park, Illinois, on July 21, 1899, and died a suicide in Ketchum, Idaho, on July 2, 1961. Probably the most influential American writer of the twentieth century, his prose style was imitated by a generation of writers, and his adventurous and highly publicized life made him known to millions who never read anything he wrote. He worked as a reporter before serving as an ambulance driver in Europe during World War I, where he was injured and decorated for heroism. After recuperating in the United States, he returned to France as a foreign correspondent and joined the expatriate literary circle in Paris that included F. Scott Fitzgerald and Gertrude Stein. In 1924, his first important work, a collection of short stories called *In Our Time*, was published. He soon emerged as a major voice of the "lost generation," writing short fiction and popular and critically acclaimed novels such as *The Sun Also Rises* and *A Farewell to Arms*. He served as a correspondent on the Loyalist side during the Spanish Civil War and fought in World War II. Much of his writing drew from events in his own life: *A Farewell to Arms* related his World War I experiences, *For Whom the Bell Tolls* (1940) concerned the Spanish Civil War, and *Death in the Afternoon* (1932) and *Green Hills of Africa* (1935) were nonfiction works about Spanish bullfighting and big-game hunting, respectively. He lived for more than 20 years in Cuba, where he wrote *The Old Man and the Sea* (1952), a novella about the courage of an aged Cuban fisherman. In 1954, he was awarded the Nobel Prize in literature.

He loved the animals he hunted, whether they were African antelopes or game fishes. In *Green Hills of Africa*, he wrote of a kudu (a species of large antelope) he had just shot:

> I looked at him, big, long-legged, a smooth gray with white stripes and the great, curling, sweeping horns, brown as walnut meats, and ivory pointed, at the big ears and the great, lovely, heavy-maned neck the white chevron between his eyes and the white of his muzzle and I stooped over to touch him to try to believe it. He was lying on the side where the bullet had gone in and there was not a mark on him and he smelled sweet and lovely like the breath of cattle and the odor of thyme after rain.

On the subject of fishing for marlin in the Gulf Stream, which he liked to call "the blue river":

> He can see the slicing wake of a fin, if he cuts toward the bait, or the rising and lowering sickle of a tail if he is traveling, or if he comes from behind he can see the bulk of him under water, great blue pectorals widespread like the wings of some huge, underwater bird, and the stripes around him like purple bands around a brown barrel, and then the sudden upthrust waggle of a bill. . . . To see that happen, to feel the fish in his rod, to feel the power and that great rush, to be a connected part of it and then to dominate it and master it and bring that fish to gaff, alone and with no one else touching rod, reel or leader, is something worth waiting many days for, sun and all.

Although *The Old Man and the Sea* is about a single fisherman and a single fish (and some hungry sharks), *Islands in the Stream* is about love, death, and war, but it is also about fishing. In this book (which was published nine years after he died), Hemingway describes in painful detail the battle between a teen-age boy and a swordfish. The boy, named David, has gone fishing in the Gulf Stream with his father, Thomas Hudson, his brothers, Andrew and Thomas Jr., and Eddy the mate. After three hours of fighting, the fish comes up:

> Then, astern of the boat and off to starboard, the calm of the ocean broke open and the great fish rose out of it, rising, shining dark blue and silver, seeming to come endlessly out of the water, unbelievable as his length and bulk rose out of the sea and into the air and seemed to hang there until he fell with a splash that drove the water up high and white.
>
> "Oh God," David said. "Did you see him?"
>
> "His sword's as long as I am," Andrew said in awe.
>
> "He's so beautiful," Tom said. "He's much better than the one I had in the dream."

The battle goes on for more than four hours. The boy's hands and feet are bloody and his back is breaking. "He was watching the fish and easing the stern onto the course that he was swimming. He could see the whole great length of him now, the great broad sword forward, the slicing dorsal fin set in his wide shoulders, and his huge tail that drove him almost without a motion . . . the fish was coming up looking as broad as a big log in the water."

"Eddy," David said. "What would he really weigh?"

"Over a thousand," Eddy told him.

"Would he really have weighed a thousand, papa?" David asked.

"I'm sure of it," Thomas Hudson answered. "I've never seen a bigger fish, either broadbill or marlin, ever."

"Well," David said with his eyes tight shut. "In the worst parts, when I was the tiredest, I couldn't tell which was him and which was me."

"I understand," Roger said.

"Then I began to love him more than anything on earth."

"You mean really love him?" Andrew asked.

"I loved him so much when I saw him, coming up that I couldn't stand it," David said, his eyes still shut. All I wanted to do was see him closer."

As the fish tires and is hauled to the surface, they are fearful that they will lose it. Eddy jumps into the water to gaff it, but "it was no good. The great fish hung there in the depth of the water where he was like a huge dark purple bird and then settled slowly. They all watched him go down, getting smaller and smaller until he was out of sight."

Maybe they don't *love* them, but it is probably fair to say that most sportfishermen respect the animals they hunt. Not everyone loves the thing he kills; it is mostly the big-game fishes (and salmon in some places) that fishermen admire so profoundly. They are, after all, only *fish*—cold-blooded, small-brained, aquatic vertebrates that sometimes fight vigorously when hooked. A game fish taking a trolled bait is probably acting out of instinct—if something is moving swiftly through the water, it must be food—which is the same instinct that causes swordfish (and marlins, sharks, turtles, dolphins, and albatrosses) to strike at a bait hanging from the gangion of a longline. I cannot find any declarations of admiration for the victims of longline fishers or commercial harpooners; to them, the fish is an object of commerce, to be

Figure 5.7. Ernest Hemingway (*left*) poses proudly with a marlin caught off Bimini. Photo courtesy of Ed Pritchard—AntiqueFishingReels.com.

dispatched as efficiently and economically as possible. The bigger the fish for the sport fisherman, the more venerable it is; the bigger it is for the commercial fisherman, the more it's worth on the dock.

The fish's acrobatics can also be used to answer the long-debated question as to whether fish feel pain. If they were not in searing agony, why else would they leap out of the water—often dozens of times in a row—and toss their heads in repeated efforts to dislodge the steel hook that has pierced their mouths? The "sport" of big-game fishing—in fact, virtually all fishing—is based on the "performance" of fish whose response to the pain of being hooked and dragged all over the ocean is to leap out of the water, toss their heads repeatedly, and sometimes dive to great depths to "escape" from the pain. In "catch-and-release" fishing, the fish puts on the requisite show for the angler, and then is rewarded by being cut loose to "fight another day." Some reward. Having expended so much energy in fighting—and being wounded in the mouth besides—released game fish often die anyway.

Michael Lerner (1891–1978), was a successful businessman who turned an early and avid interest in hunting and fishing into a mature scientific avocation. Michael was one of seven children of Charles and Sophie Lerner. With his father and brothers, he founded a nationwide chain of women's clothing shops, known as Lerner Stores, but he left the business after it had provided enough money for him to pursue his passion for catching very large fishes.[5] Michael and Helen Lerner visited the island of Bimini in the 1930s, and by 1936, they had founded the Bahamas Marlin and Tuna Club, with Ernest Hemingway as the first president. Around this time, Lerner's association with the AMNH began, as he would donate his prize catches, often mounted in lifelike poses, to become museum exhibits. He pioneered the rod-and-reel fishery for giant tuna in 1935, in Wedgeport, Nova Scotia, and as a trustee, he led seven American Museum expeditions (always where there was good fishing) to remote locations such as Cape Breton Island (Nova Scotia), Bimini, Australia, New Zealand, Peru, Chile, and Ecuador.

In 1935, Lerner fished successfully for giant tuna off Wedgeport, but when he learned that commercial swordfishermen were harpooning as many as 150 fish per year off Cape Breton Island (the eastern end of

5. The same retail business provided the money for lyricist Alan Jay Lerner (1918–86), Michael's cousin. Alan also did not follow in his father's footsteps, but pursued a career in music, first contributing to the Hasty Pudding shows at Harvard in 1938 and 1939 and later studying at Julliard. In 1942 he met composer Frederick Loewe, and they began their collaboration on such hit musicals as *Brigadoon* (1947), *My Fair Lady* (1956), and *Gigi* (1958). Lerner also wrote the screenplay for *An American in Paris* (1951). He rejoined Loewe for *Camelot* (1960) but they had a falling-out and went their own ways.

Nova Scotia), he raised anchor and headed for Louisburg , Nova Scotia. Called "an important swordfish fishery" by J. T. Nichols and Francesca LaMonte in their 1937 paper, Louisburg could boast over 300 fish being brought in by fishermen on a good day, the dressed weight averaging 265 pounds. Examination of the stomach contents revealed that the favorite food of Cape Breton swordfish was herring, but there were also squid, mackerel, and in one, a whole dogfish (*Squalus*). With Captain Tommy Gifford, whom he brought with him from Bimini, Lerner boated a 426-pound swordfish, the first rod-and-reel swordfish ever taken off Nova Scotia. As quoted by Mike Rivkin, Lerner wrote a letter to Hemingway in which he gushed that "the ocean was covered with swordfish . . . more than 60% weighed 500 lb. or more." At Cape Breton Island, he landed several more large swordfish, one of them weighing more than 600 pounds. Francesca LaMonte of the American Museum of Natural History was one of the ichthyologists on this expedition, and in her children's book, *Giant Fishes of the Open Sea*, she described Lerner's fishing style at Louisburg:

> He was accustomed to using the best equipment, including a well-outfitted cabin cruiser, but here at Louisburg, things were different. He had to do his fishing from a small dory. It was an awkward and dangerous arrangement. His swivel fishing-chair had to be crudely attached to the bottom planking of the boat. This clumsy, low-lying craft pitched and rolled in the rough water, but Mr. Lerner expertly handled his rod and reel and made some very fine catches.[6]

According to Gudger, Michael Lerner brought back 141 swords (but not their original owners) from this expedition. (My efforts to track down the swords in the museum's collection proved fruitless—or more

6. LaMonte also wrote (of the commercial swordfishery of Louisburg): "When the fishermen come in with their catches, the huge swordfishes, still glistening with salt water, are lifted onto the docks. . . . The cleaned and gutted swordfishes then are loaded on a dolly and pushed to a weighing shed. . . . The livers of the swordfishes are gathered up and packed into large cans to be sent to chemical companies which use them for their vitamin content. The swordfish hearts are also collected, these by little boys who impale them on metal rods and carry them home to be cooked and eaten."

precisely, *swordless.*) In 1938, his wife Helen was the first woman to catch a broadbill off Nova Scotia a 320-pounder, and she later caught one off Chile, thus becoming the first woman ever to land a swordfish in the Atlantic and the Pacific. Off Chile in 1939, according to his sometime fishing companion Kip Farrington, Lerner "pulled the greatest thrill and the hardest thing to do in salt-water angling—taking two swordfish in a single day."

Lerner researched and recorded migration patterns of marlin and swordfish throughout the oceans and was the principal organizer of the International Game Fish Association, which held is first meeting at the American Museum of Natural History on June 7, 1939. Also in

Figure 5.8. By the mid-twentieth century, swordfishing had become a significant part of Canada's maritime economy. Swordfishing boats at Glace Bay, Nova Scotia Bureau of Information, photographer, ca. 1948; NSA Photograph Collection: Places: Glace Breton: Glace Bay: Swordfishing boats (NSIS neg. no. 3796).

attendance were big-game fisherman Van Campen Heilner and museum ichthyologists Francesca LaMonte, John Nichols, and William Gregory. Michael Lerner agreed to "personally underwrite all the expenses of the new organization, a level of support he would continue to maintain until his resignation as IGFA president in 1961. He also prevailed on celebrated wildlife painter Lynn Bogue Hunt to design the first IGFA logo, which consisted of a swordfish chasing tuna. (In 1996, IGFA retired the first logo and replaced it with a multispecies version created by Guy Harvey.) For those who would track Mike Lerner's not insignificant accomplishments as an angler and organizer of expeditions and fishermen, it is disappointing that Mike Lerner never wrote a book, leaving the pelagic game fish chronicles to writer/fishermen Zane Grey and Ernest Hemingway and fishermen like Kip Farrington, Ralph Bandini, and Philip Wylie.

According to Mike Rivkin's history of IGFA, Philip Wylie (1902–71) was "the hardest working of the original IGFA angler/founders." Using the fictional "Crunch and Des" as protagonists, he described adventures of fishing for sailfish, marlin, swordfish, tarpon, barracuda, sharks, and giant jewfish in the Gulf Stream and the Gulf of Mexico. Wylie's immensely popular stories about saltwater fishing are filled with excitement, the thrill of marine sport, and shrewd character studies. In her introduction to the 2002 reissue of the Crunch and Des stories, the

Figure 5.9. Swordfish on IGFA's first logo, designed by Lynn Bogue Hunt. International Game Fish Association.

author's daughter, Karen Wylie Pryor, wrote: "For the reader about to breathe Wylie's velvet air, hear the chug of charter boat engines, and cross the bridge of indigo-blue of the Gulf Stream for the first time—I envy you." Here's Karen's dad writing about fighting a broadbill:

> And soon, a hundred yards away, the ocean split open. It tossed forth a biological dynamo, a blue and silver being, upward of ten feet in length, with a nose sword as wide as a man's hand and nearly a yard long. This beast seemed to ride on air, leaving a brief trough in the water. It rolled on one side and then the other. It bent itself in almost U-shape. It flexed straight and re-entered the sea, bill first, neat as a diver. It came right back out, as if bouncing, and tail walked in a great arc that brought it closer to the boat.

Mike Lerner had been officially appointed a field associate of the AMNH, which meant that he maintained a professional affiliation with the museum and donated many of his catches—specifically game fish and squid—to the collections, some of which were mounted and proudly displayed in the museum's exhibition halls. Eventually, the Lerners' home and office on North Bimini Island was turned into a full-scale marine laboratory and functioned as an adjunct facility of the museum's Department of Fishes and Aquatic Biology. The laboratory provided researchers with living accommodations, boats, laboratory equipment, indoor aquaria, and outdoor pens for the study of sharks, sawfish, tarpon and other large fish. The Lerner Marine Lab was closed in 1977 for lack of funding, and the Lerner-Gray Scholarship Fund was set up in its place.

Like every other international big-game fisherman, Mike Lerner was more than a little interested in records. The official world's record swordfish—the 1,182-pounder—was caught by Lou Marron off the coast of Iquique, Chile, on May 7, 1953. Eugenie Marron wrote a book about her husband's (and her own) swordfishing exploits off Chile, but also in Nova Scotia waters and the Humboldt Current, where they reeled in those ink-squirting jumbo squid. Eugenie was no slouch as an angler herself; she succeeded in boating a 772-pound swordfish and titled her book *Albacora*, which she tells us is the "Spanish idiom for broadbill swordfish." The book pays homage to her husband, "Uncle Lou" Marron, (chairman of the board of the Coastal Oil Company of

New Jersey), but even more to the great and wonderful swordfish, which the Marrons liked to refer to as Bosco. Here is Eugenie's account of her husband's record-breaking accomplishment:

> "Bosco!" Lou bellowed at once. . . . Everyone aboard jumped into action, and baiting began before the sound of Lou's shout faded. Lou pulled line off the big Fin-Nor reel and stood holding the end of the big, trailing loop in his hand. We waited, breathless. The huge dorsal fin looked like the mainsail of a windjammer. Eddie Wall, the veteran of many swordfish safaris, was our captain that year. He put that bait right across the albacora's nose. The fish turned away. Again we maneuvered. This time with a great splash of his mighty tail, Bosco dove for the bait. "Wham"—the sword lashed out. Lou was in the chair now, waiting and watching. . . . A little of the line peeled off, then a little more, faster and faster. . . . We had struck Bosco, the king of kings. Lou worked like a man driven by some fury risen from the sea. It was scarcely more than fifty-five minutes before that greatest of all albacora was brought to the leader. The next hour taxed us almost beyond endurance: twelve times more, twelve separate battles, each one bringing him to the boat, only to have him elude our grasp and dash away. . . . Eddie wrapped the heavy-cable leader wire around both hands and held on. I thought his arms would rip from his shoulders—but the gaff was in the fish now and he was ours—Bosco, the king of the sea! What a Goliath!

Surely a female, "Bosco, the king of the sea," was the largest game fish of any species that had been caught on rod and reel anywhere in world up to that time. With the proud conqueror alongside, the fish's obituary picture graces the pages of *Albacora*. (The fish was brought back to the United States and a mount now hangs in the clubhouse of the Miami Beach Rod and Reel Club.) A swordfish hung by the tail gives only a hint of its massive power when alive. You notice its beautifully streamlined shape, its crescent tail, and its huge eyes, but it is the sword, often a third of fish's total length, that makes you realize that *Xiphias gladius* is something very special indeed. She is the *queen* of the sea, slasher of the depths and of the surface, boat-piercer, man-chaser. She is a fish not to be taken lightly.

Swordfish repeatedly demonstrate their unwillingness to be caught.

Figure 5.10. At 1,182 pounds, Lou Marron's prize is still the largest swordfish ever caught. Iquique, Chile, 1935. International Game Fish Association.

They drive their swords through dories; they dive madly toward the bottom when harpooned; and when hooked, they leap and fight as if possessed. And if you thought that a swordfish on a longline hook might go quietly, read Captain Linda Greenlaw's description of her attempt to bring a really big swordfish aboard the *Hannah Boden*:

Figure 5.11. Eugenie Marron, shown here with her 772-pound swordfish caught off Iquique, Chile, in 1954, had no idea that all the big ones were females. International Game Fish Association.

> I recalled one gigantic warrior of a sword that made the mistake of taking the bait, but refused to let the mistake become fatal. . . . I fought this monster with 100 feet of steel boat and the hydraulic spool for forty minutes without so much as seeing the snap of the leader that held him. . . . When the heavy snap finally appeared, I thought the battle was over, and it was, but the war had just begun.

> I fought the fish with arms, legs, back, shoulders, everything I had, never gaining more than 2 fathoms before the monster greedily took them back from me. After a few minutes of tug-of-war, I allowed one of the crew to take over for me, and another man relieved him, also becoming exhausted, and we still had not seen the fish. . . . The fish had finally given up, I thought, and was treading water at my feet just below the surface. It was by far the largest swordfish I had ever seen. I anxiously waited for the gaffs to be driven deeply into its massive head. My blood surged with excitement and I was intoxicated with the size of the fish now within our reach.

The fish surfaces, looks at Greenlaw with its "grapefruit-size eye," and then dives. She attempts to hold it by bending the wire leader over the rail, thinking that the fish would now have to pull against the entire boat, not just a couple of puny fishermen. The line parts and the fish escapes, but almost immediately returns to the surface "gloating in its freedom and rubbing my face in defeat" it takes its "victory lap." This so enrages Greenlaw that she joins her crew members in screaming violent expletives as the victorious swordfish swims away. She writes, "I am afraid that I have never had much of that 'Free Willy' spirit, and confess that I have only bitter feelings about the one that got away." If this had been sport fishing, the fisherman who had lost the monster after a mighty battle might have saluted it, but there is no sport in long-lining, only money to be made or lost.

Though swordfish are still considered premier game fishes and are eagerly sought as trophies, the large ones are no longer available in the North Atlantic. In the past, swordfishermen hunted their prey by sight, waiting for the telltale sickle-shaped dorsal fin and upper lobe of the tail to break the surface as the big fish "finned out." As Graham Faiella wrote in *Fishing in Bermuda*, "They feed nocturnally on squid and surface apparently to aid digestion in warmer water . . . swordfish on the surface, having eaten their fill rarely take an interest in a surface trolled bait." Swordfishermen would have to harpoon the fish from a specially designed "pulpit" that extended from the fishing boat's prow. The harpoon fishery for the Atlantic swordfishery was limited to the New England coast in the early 1800s. When Jordan and Evermann wrote *American Food and Game Fishes* in 1902, they said this about the swordfish: "The species is rather abundant for so large a fish. Off the New England Coast, 3,000 to 6,000 of these fish are taken every year. Twenty-five or more are sometimes seen in a single day. One fishermen killed 108 in one year." "Them days," as the fishermen say, "is gone forever."

In May 2005, in the sport fishing magazine *Marlin*, Mike Leech described the history of swordfishing tournaments in Florida. The first such tournament was held in Fort Lauderdale in 1977, "with spectacular results," but as long-liners descended en masse to the Straits of Florida—a known breeding ground for *Xiphias*—the catches and the average size of the fish dropped dramatically. By 1983, when no catches and no strikes were reported, the tournament was called off, but it was resumed in 2002, when it was believed that the stocks had recovered. In that year, the average swordfish caught weighed 110 pounds; the next year it fell to 93 pounds, and by 2004 it was 82.5. These weights are well below what a female swordfish weighs before her first spawning, so removing immature females from the gene pool did not benefit the overall population. Because sport fishermen account for only one-half of 1 percent of the US swordfish quota (8,595,400 pounds in 2004), the population crash and diminution in size of the swordfish caught is almost solely because of long-lining. "Florida's shrinking swordfish may or may not indicate an Atlanticwide trend," says Leech; "only time will tell. But if the average size of landed swordfish continues to fall, it may indicate a serious problem with the spawning stock biomass of the species. The Straits of Florida are a proven spawning ground. If only a small percentage of fish reach spawning size, then the so-called rebounding swordfish population may take a turn for the worse."

Swordfish have been targeted by commercial fishers for centuries. The early fishermen who used harpoons and spears had a deleterious effect on individual swordfish but were of little consequence to swordfish populations as a whole. It was not until the introduction of longlines, unquestionably the deadliest (and most effective) fishing method ever devised, that swordfish populations began to tumble. Here is the 1998 description of the longline fishery for swordfish by Carl Safina:

> Today, the submarine canyons and banks these animals prowl is so spaghettied with baited lines that more than 80 percent of the fe-

male swordfish caught are immature, killed before they can breed. (Female swordfish take at least five years to reach sexual maturity, at which point they are almost six feet long.) With longlines taking 98 percent of the swordfish catch, large adult fish are rare in the North Atlantic. Between the early 1960s and today, the average size of North Atlantic swordfish dropped two-thirds, from almost 270 pounds to 90 pounds. In nursery grounds off Florida, off South Carolina, in the Gulf of Mexico, and elsewhere, longlines catch mostly juveniles.

All the spear-nosed fishes are related, but the swordfish, the only one with a flattened sword, has been put in a class by itself. Zane Grey, Ernest Hemingway, Kip Farrington, Michael Lerner, and Philip Wylie admired the swordfish almost to the point of reverence, but with the partial exception of Grey's *Tales of Swordfish and Tuna*, no one ever devoted a whole book to *Xiphias gladius*. The swordfish stands alone in so many areas: close relatives (it has none); diving ability (it has no equal); fighting ability (ditto); aggressiveness (double ditto); overfishing (with the tunas, it is among the most heavily harvested large fish species in the world). These qualities, together with its powerful, streamlined beauty, make it the functional exemplar of the term "sui generis." If ever a fish deserved a book of its own, it is this one.

Actually, *Xiphias gladius* already has a book of its own. In 1942, Kip Farrington, the famed big-game fishermen whose exploits were discussed earlier here, wrote *Bill, the Broadbill Swordfish*, which was illustrated by Lynn Bogue Hunt. "Bill" is a young swordfish, born in the Mediterranean, who refuses to eat anything that could contain a baited hook. He travels the world with his father and mother until she is killed by a mako shark and his father is caught by a fisherman off Montauk. He meets Albacora, the love of his life, off Tocopilla, Chile, and he now weighs 1,300 pounds. He is still too smart to take a baited hook, so he and Albacora will live happily ever after: "Bill has no fear of anglers. All sharks are afraid of him. . . . because of his huge size and his long and dangerous sword." As with most anglers of the period, Farrington believed that the really big swordfish were males, which was completely wrong.

Fish do not have heels, but if they did, the great marlins would be

Figure 5.12. In 1942, big-game fisherman Kip Farrington wrote *Bill, the Broadbill Swordfish*, illustrated by Lynn Bogue Hunt. From *Bill, the Broadbill Swordfish* (1942).

Figure 5.13. When Farrington wrote *Bill, the Broadbill Swordfish*, everybody believed that males were bigger than females. They were wrong. From *Bill, the Broadbill Swordfish* (1942).

Figure 5.14. To honor the most celebrated of all big-game fishes, the International Game Fish Association commissioned Kent Ullberg's *Sword Dancer* for their headquarters in Dania Beach, Florida. International Game Fish Association.

hot on those of the swordfish. The blue marlin and the black are also giants that fight spectacularly when hooked. Fishermen often travel thousands of miles—to Hawaii, New Zealand, Australia, Chile, Costa Rica, Panama, Baja California, the Bahamas, West Africa—for the opportunity to fight a 1,000-pound marlin for hours in the hot sun, strapped into a chair bolted to the deck, as the strain threatens to pull their arms from their sockets. The smaller billfishes—the white and striped marlins, the spearfishes and sailfishes—are also zealously sought around the world. Perhaps it is the rapier-like anterior extension that elevates these sleek torpedoes to a different level of piscine beauty (although, even lacking a sword, the tunas are also considered beautiful), but whatever it is, the billfishes are the reigning royalty of marine game fishes.

6 The Sword's Relatives

Billfish are the superfish, even those that do not rate the name "giant." . . . Their color, conformation and power are a tribute to nature's evolution. . . . The unresolved answers to many questions about them are sought by scientists, fishermen and fishery managers alike. Billfish have gripped the imagination and triggered our quest for knowledge as have no other fish. They are the stars of the fishing world.—Peter Goadby, *Billfishing*

In 1857, Albert C. L. G. Günther, born in Germany in 1830, was appointed keeper of the Zoological Department of the British Museum of Natural History, where his responsibilities included recording the multitude of specimens in the museum's collection. Toward that end, he produced the eight-volume *Catalogue of the Fishes in the British Museum*, a volume of which was published every couple of years from 1859 to 1870 and included descriptions of every species in the collection. In 1880, he wrote *Introduction to the Study of Fishes*, which included the history of ichthyology, chapters on the anatomy, distribution, and paleontology of fishes, and 400 pages on the known families. An ichthyologist with the resources of one of the world's great museums at his disposal, Günther had this to say about the "Sword-Fishes":

> The upper jaw is produced into a long cuneiform weapon. These fishes form one small family only, *Xiphiidae*. The "Sword-fishes" are pelagic fishes, occurring in all tropical and sub-tropical seas. Generally found in the open ocean, always vigilant, and endowed with extraordinary strength and velocity, they are but

> rarely captured, and still more rarely preserved. . . . The distinction of the species is beset with great difficulties, owing to the circumstance that but few examples exist in museums, and further, because the form of the dorsal fin, the length of the ventrals, the shape and length of the sword, appear to change according to the age of the individual. Some specimens or species have only the anterior dorsal rays elevated, the remainder of the fin being very low, whilst in others all the rays are exceedingly elongate, so that the fin, when erected, projects beyond the surface of the water. It is stated that Sword-fishes, when quietly floating with the dorsal fin erect, can sail before the wind, like a boat.

"Cuneiform" means wedge-shaped and is commonly used when describing the incised, arrowhead-shaped strokes used in systems of ancient Assyrian or Persian writing. Otherwise, Günther is telling us that at that time, nobody knew very much about swordfishes, and the sailfish—which "can sail before the wind, like a boat"—had either not been differentiated as a separate species or, more likely, had not made it into the British Museum's collection. And except for some vague references to different shapes of pectoral fins, swords, and dorsal fins, nothing that might be a marlin had been recognized as of 1880.

Although the swordfish has been known since antiquity, its living relatives, now listed among the most popular game fishes in the world, have revealed their existence to science only recently. Museum fish collections usually take the form of specimens in jars, and it is more than a little difficult to imagine the jar that would be required to hold a full-grown swordfish. Even if there are billfishes in the collection, they are likely to be juveniles that fit in jars, and which, as Günther pointed out, change dramatically on their way to adulthood. Because a "growth series" is usually lacking, it has proven to be very difficult to distinguish the juveniles of the various species. While there may indeed have been the odd adult marlin or sailfish accidentally snared in a net and brought to the nearest museum (if there was one), the swordfish's relatives became known only in modern times—mostly because they were caught for sport. And as we have seen, sport fishing is a rather recent innovation in the long history of man versus fish.

With the introduction of marlin fishing off Chile and bluefin tuna and swordfishing off California in the early decades of the twentieth century, interest in the large spear-nosed fishes began to gather momentum, and sportsmen recognized that there was more to fishing than hauling in a fish so you could eat it. Zane Grey's enormously popular books on fishing for tuna, swordfish, and marlins spread the word about the excitement entailed in battling these fighting giants, and today, fishing for billfishes is the ne plus ultra of big-game fishing. No other species are so elusive, fight so hard, or look so good over the fireplace—and very few species have their own magazine. In the magazine called *Marlin*, you will find spectacular photographs of leaping billfishes (mostly the magazine's eponym), articles about where and how to catch billfishes, accounts of important moments in the history of billfishing, profiles of great fishermen, and advertisements for everything related to billfishing, including rods, reels, baits, lures, clothing, sunglasses, resorts, and charters. But if there is one thing that attests to the immense popularity of the sport, it is the voluminous pages of ads for boats. Leafing through the magazine, you would quickly arrive at the conclusion—which is largely correct—that big-game fishing is the special province of people with hundreds of thousands of dollars to spend on a 45- (or 60- or 75-) foot, fully equipped fishing boat. Like the prey they are designed to hunt, these boats are sleek, fast, and powerful; and if the fish wins the battle—that is, if the fish escapes—it demonstrates that the opponents were (almost) equally matched.

There is a increasing tendency among fishermen to "catch and release" fish, so that the fish—which probably would not be eaten anyway—does not have to die for the momentary triumph of the fisherman or the trophy in the game room. And it turns out that you don't even need the fish to display it in your home because the fish over the fireplace is hardly ever the fish that was caught. Skinning and mounting a fish is a complicated and expensive process, and "skin mounts" don't last, as the skin soon begins to peel away from the mount. (Besides, what are you going to do with an 800-pound fish that you caught in New Zealand? Mail it to a taxidermist in Florida?) Nowadays, most triumphant anglers record their catch with a photograph and then order a fiberglass replica of the fish they caught, measured to the inch and painted in lifelike colors. Some taxidermists will make a mold of the actual fish and from that make a fiberglass replica, but for the most part, fishes over the fireplace are made of plastic.

Even though the marlins are highly respected as game fishes, they are not held in such high regard by commercial fishers, and for longliners, where tunas and swordfish are the highest-ticket items, blue and white marlins are considered "bycatch"—the euphemism for undesirable species caught in a given fishery—and are discarded. In a 2004 study of the Japanese pelagic longline fishery in the Atlantic from 1960 to 2000, Serafy et al. found that, "for at least the last two decades, the principal source of mortality on adults of both Atlantic marlin species has been pelagic longline fishing. This method of fishing deploys a continuous mainline, up to 60 miles in length, with regularly spaced branch lines which terminate with baited hooks." In the 1960s, when the fishery began, the Japanese deployed millions of hooks in the Atlantic and caught millions of tons of albacore (by far the predominant species), yellowfin and bigeye tuna, and proportionally fewer tons of swordfish. In 1962, 340,000 white and blue marlins were caught and discarded; the 1964 totals for both species were somewhat lower. For target and nontarget species the numbers began to decline after that, even though the number of hooks remained high. In their article, "Rapid Worldwide Depletion of Predatory Fish Communities," fisheries biologists Ransom Myers and Boris Worm looked at the same data from the Japanese longline fishery that "represent the complete catch-rate for tuna (Thunnini), billfishes (Istiophoridae) and swordfish (Xiphiidae) aggregated in monthly intervals from 1952 to 1999, across a global 5° × 5° grid" and estimated that only 10 percent of the biomass of these species remains. They concluded that "blue marlin was initially the dominant billfish species but declined rapidly . . . probably due to increased 'bycatch' mortality," so even those species that are not targeted and considered "incidental" are in a population free-fall. It seems that it is uneconomical to bring in the Atlantic marlins, so even though they are edible, they are discarded. What a waste.

Marlin sushi (*makajiki*) is popular in Japanese restaurants (particularly in Japan), and there are some regions that serve marlin steaks, often anonymously, as "catch of the day." In Hawaii, the blue marlin (known by the Hawaiian name *a'u* or the Japanese name *kajiki*) is a popular target for sport fishermen—particularly out of the Big Island port of Kona—and fish brought to the dock are often sold to fish wholesalers or restaurants. The striped marlin (*nairagi*) is usually much smaller than the blue and is caught in Hawaiian waters by long-liners or sport fishermen. Also caught by long-liners is the shortbill spearfish, known in Hawaii as *hebi*. The mild, amber flesh is sold throughout the year at Honolulu fish markets, but because restaurateurs and their customers are not familiar with this fish, it is often called something else, or simply described as billfish. Long-liners throughout the world's oceans often haul in marlins of any and all species. Unlike tuna or swordfish, marlin is not a particularly popular menu item, so it is chopped up and used in dip, stew, or any other dish requiring fish parts. There are some areas, however, where marlin are caught, but not eaten. In *Fishing in Bermuda*, Graham Faiella notes that

> there is no particular reason for landing marlin in Bermuda apart from showing the fish as a trophy. There is virtually no local demand for it as a food fish unlike in the Caribbean and other parts of the world where it provides an important food source. Its dark, oily flesh is not to Bermuda tastes (unless cured like smoked salmon), accustomed as local customers are to the more refined flavors of tuna, wahoo, snapper, and other lighter flesh fish.

Even though it not fancied as a food fish, the blue marlin is Bermuda's premier big-game fish. Faiella called *Makaira nigricans* "the largest and most spectacular gamefish regularly fished in Atlantic waters." Indeed, with its even larger "black" relative, the blue marlin is among the most spectacular game fishes in the world.

THE BLACK AND BLUE MARLINS

Size, beauty, and a ferocious, acrobatic reluctance to be reeled in are the qualities that define the billfishes as the world's most important game fishes. They are all large—some are among the largest and heaviest of all fishes—but it is their matchless jumping and fighting ability that makes them especially worthy opponents for fisherman. Yes, a hooked

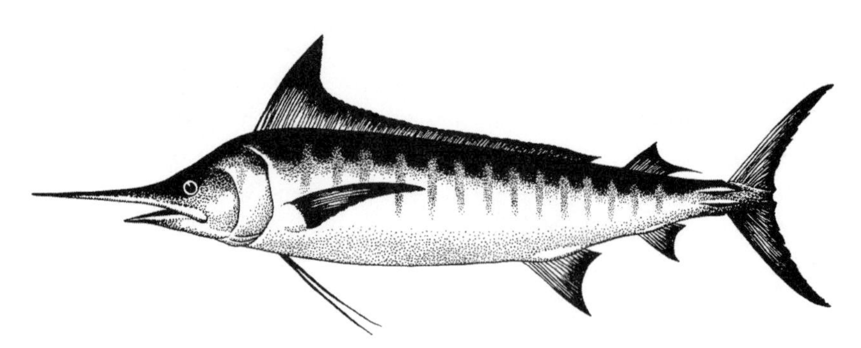

Figure 6.1. Blue marlin. Drawing by Richard Ellis.

tarpon or salmon will leap enthusiastically, but a 1,000-pound marlin tail-walking or "greyhounding" (leaping out of the water in repeated low-arc jumps) represents the ultimate thrill for the deep-sea angler. Some species are known for their aggressiveness—the marlins are second only to the swordfish in the number of attacks on boats—and divers in marlin-infested waters know enough to steer clear of billfish. In a 2005 article about aggregating fish species in the vicinity of Gulf of California seamounts, scientist/divers Peter Klimley, John Richert, and Salvador Jorgernsen wrote,

> Four species, three marlins and a sailfish, frequent the seamount during the day. . . . Striped marlin and sailfish frequent [the seamount] from spring through fall, but the larger, rarer blue and black marlins usually arrive in late summer. They tend to loiter close to the surface, maintaining their position by slowly sculling with their pectoral fins. They also tend to be unfriendly: On encountering a diver, a billfish will often put on an aggressive display by extending its fins to appear bigger, opening and closing its jaws and shaking its rostrum to emphasize its formidable armament. At this point it is prudent for a diver to swim away at the maximum angle of escape to avoid a charge from the fierce fish.

But when piscine painter Stanley Meltzoff wanted to see one of these "fierce fish" up close and personal, he joined a fishing expedition out of Saint Thomas in the Virgin Islands with the intention of observing a marlin in the water. In a 1977 *Sports Illustrated* article titled "Like a Neon Shadow in the Sea," he described his first sighting:

> On first seeing a blue marlin, as at the first sight of a rhinoceros or perhaps more apt, if less possible, a unicorn, there seems very little to say besides some four-letter variant of "gee whiz." The external oddities of this intimidating creature are but the beginnings of its marvels. Its stiffened pectoral fins serve as ailerons for banking, but they also swivel in their sockets to slow down the marlin, like the speed brakes on a jet. Two spines, remnants of the ventral fins, are kept tucked into a much longer slit that runs the length of the fish's belly. No one has ever seen these extended in use or has offered a possible explanation of their function.

Meltzoff actually gets in the water with the great fish ("the blue marlin regarded us reflectively for a moment and beckoned us with its sword glowing dull red and glinting bronze. Sheathed in midnight, the fish glided down into the profound blue and we followed") but decides that "the blue marlin in the sea is better shown than told" and accompanies his article with several spectacular paintings. In a 1988 *Field and Stream* article about Meltzoff, Duncan Barnes quotes the artist's first impressions: "It had a head like a grand piano and a pike as long and thick as a baseball bat and as long as a man. The head widened into shoulders as broad as a bison's. The marlin terminated behind the erect dorsal fin; the rest had been betten off by sharks—scalloped tooth marks were clearly outlined. I was ready to quit before I started."

Guy Harvey has become renowned for his paintings of big-game fish and also for his willingness to jump in the water to photograph them. His book *Portraits from the Deep* is lavishly illustrated with his paintings and his underwater photographs of marlins, sailfish, and swordfish. (That's his photograph of a swordfish on p. 32.) On diving with a big blue marlin in the Cape Verde Islands, he wrote, "I dropped in and had a great angle on a 300-pound blue that hung around for a short time. Her body was completely black, almost like a Pacific sailfish's: no stripes, no blue tail—just a black missile. She didn't participate in the teasing process, and as I followed her away, up-sea, she retuned to her normal pastel-blue cruising colors." While size and range are useful for differentiating the marlin species, the fish themselves can lose or brighten their stripes and even change color completely. Marlins that are brought aboard a boat fade quickly, as does the dolphinfish,

Coryphaena hippurus, celebrated for losing its bright blue, green, and yellow coloration as it dies. In *Childe Harold's Pilgrimage*, with the fish rather than the marine mammal in mind, Byron writes:

> Parting day
> Dies like the dolphin, whom each pang imbues
> With a new color as it gasps away,
> The last still loveliest, 'till –'tis gone,
> And all is gray.

Fish are well-known for their ability to change color. Think of the flatfishes that can modify their dorsal coloration to match their surroundings, or some of the wrasses and gobies whose daytime coloration is totally different from their nocturnal raiment. As in many species of cephalopods—even more adept at color modification than fishes—the pigment in the billfishes' skin is contained in cells called chromatophores, which can be expanded or contracted and thus change the external appearance of the fish. One type of chromatophore is the iridophore, which contains guanine and purine, pigments that are responsible for the colors white, silver, and blue that are characteristic of the billfishes. In a sailfish or marlin, the stratum argenteum, a sheet made up of layers of guanine-containing iridophores, reflects and scatters most of the light that hits it, making the light appear silvery. If the crystals or iridophores are aligned in such a way that the light waves are reflected back in parallel, it appears iridescent. The epidermis of many fish species also contains iridocytes, reflective cells that contribute to the silvery sheen, iridescence, or "neon" flashing of some species. As Peter Moyle and Joseph Cech put it (in their *Fishes: An Introduction to Ichthyology*), "Adrenalin and its relatives cause pigment granules in the chromatophore to aggregate, which rapidly makes the fish lighter in color, whereas acetylcholine makes them disperse slowly. Thus fish that are frightened may undergo rapid color changes and then slowly revert to their original colors when danger is past." The marlins are able to initiate color changes that are nothing short of spectacular.

Like Guy Harvey who glorifies them on canvas, Ernest Hemingway believed that the marlins were among the most beautiful creatures on earth. In *The Old Man and the Sea*, the old man has hooked a great fish that he is trying to bring in:

> The line rose slowly and steadily and then the surface of the ocean bulged ahead of the boat and the fish came out. He came out unendingly and water poured from his sides. He was bright in the sun and his head and back were bright purple and in the sun the stripes on his sides showed wide and a light lavender. His sword was as long as a baseball bat and tapered like a rapier and he rose his full length from the water and then re-entered it like a diver and the old man saw the great scythe-blade of his tail go under and the line commenced to race out.

Because the old man's fish was caught in the Gulf Stream, it must have been an Atlantic blue marlin, whose scientific name is *Makaira nigricans*. (*Makaira* comes from the Latin *machaera*, which means sword; and *nigricans* which means black or blackish; so the blue marlin is "blackish," an example of the confusion in identifying the various species by their common names.) Except for a difference in the lateral line (a visible line of pressure-sensitive cells that runs from gills to tail) it is the same fish as the Indo-Pacific blue marlin, *Makaira mazara*, and many taxonomists do not believe that the lateral-line differentiation is significant enough to separate the two species. Their habits, size, and coloration are the same; *M. nigricans* of the chicken-wire-shaped lateral line is found in the Atlantic; *M. mazara* of the simple-looped line is found throughout the Pacific, from the west coasts of North and South America all the way to the Indian Ocean shores of East Africa. The black marlin, *Makaira indica*, is found only in the Pacific and, unlike most of its relatives, usually has no stripes.

The old man's fish—or what's left of the carcass after the sharks have demolished it—is "eighteen feet from nose to tail." According to Nakamura, "the maximum size of this species exceeds 3.75 m [12.3 ft.]," but because most record fishes are weighed and not measured, there are few authoritative length records. I have been unable to find any blue marlin longer than 15 feet, and the 15-footer, like the old man's 18-footer, is either a guess or an exaggeration. The official record weight for *M. nigricans* is 1,402 pounds; for *M. mazara*, 1,376, although IGFA does not recognize *M. mazara* as a distinct species and only distinguishes them geographically, calling both the Atlantic blue marlin and the Indo-Pacific blue marlin *Makaira nigricans*. Peter Goadby writes that the Indo-Pacific version can reach 2,600 pounds, and the Atlantic, 2,000, but no fish of either species weighing that much has ever been caught with rod and reel. The Pacific blue marlin known as Choy's Monster, caught in Hawaiian waters in 1970, weighed 1,805 pounds, and is the largest billfish ever caught by an angler, but more than one person handled the rod, so it is ineligible for the IGFA record book.

In the IGFA newsletter *International Angler* for January–February 1995, there is a photograph of a gigantic blue marlin that was caught off Portugal on July 25, 1993, with the weight given as 1,741 pounds. It seems that the wife of the captain of the *Rascasco* sent the measurements to IGFA and asked them to estimate the marlin's weight since the scale had broken when the fish was hoisted up. The fish was cut up

Figure 6.2. The world's record: Alfred Glassell's black marlin caught off Cabo Blanco, Peru, in 1953. Not clearly visible on the side of the fish is the weight: 1,560 pounds. International Game Fish Association.

the night it was caught and was never weighed, but the captain submitted the fish for an all-tackle record some months later claiming it was later weighed on another scale. With nothing but the captain's word and the photograph, the record application was rejected. (This was the application that was accompanied by the two broken-off bill tips mentioned in chap. 4.) In *Tales of Tahitian Waters*, Zane Grey looks at his 1,040-pound marlin, which would have weighed a couple of hundred pounds more if the sharks hadn't gnawed on it, and "could not help but remember the giant marlin Captain had lost in 1928, which we estimated at twenty-two or twenty-three feet, or the twenty-foot one I had raised at Tautira, or the twenty-eight foot one the natives had seen repeatedly alongside their canoes." A 28-foot marlin—if such a fish existed—would probably weigh more than a hippopotamus and would have been more than a match for any angler, even the redoubtable Zane Grey.

Just as the swordfish gets its name from the resemblance of its upper jaw to a sword, the marlins get their common name from the marlinspike, a pointed iron tool from the days of sail that was used to separate strands of rope—sometimes known as marlines—in splicing. (There are even some references to "marlinspike swordfishes.") There may be three species of marlin—or four, or five, or 16. The confusion was exacerbated by big-game fishing writers like Zane Grey and Hemingway, who used different common names for different species; Grey liked to call marlins "marlin swordfish," for example, and it is likely that neither of them knew (or cared) how many large marlin species there were, and they paid only passing attention to scientific nomenclature. In Grey's *Tales of Tahitian Waters*, we find this mystifying sentence: "When we arrived at our dock we pulled the swordfish ashore and strung him up. He was a superb specimen of the ordinary striped marlin spearfish." In 1930, Grey decided to name his 1,040-pounder a "Giant Tahitian Striped Marlin," but in a 1935 note, J. T. Nichols and Francesca LaMonte confounded the nomenclature even further, as they decided that "the Tahitian black marlin, or silver marlin swordfish," caught by Eastham Guild in 1931 (the photograph sent to them by Zane Grey), was "a recognizable undescribed form."

The actual number of marlin species, and for that matter, how to tell them apart, has perplexed ichthyologists and fishermen for years.

In another 1935 article titled "How Many Marlins Are There?" Nichols and LaMonte tried to clarify the situation: "Some persons interested in the subject," they wrote, "believe that there is but a single true marlin, and that the various forms represent age, sex, and individual variation. On the other hand, Jordan and Evermann . . . in 1926, recognize sixteen species of Makaira."[1] Nichols and LaMonte concluded ("subject to revision on the receipt of more data") that there are "three well-marked species of *Makaira,* each with a wide range, and very likely with recognizable geographic varieties in different localities." They identified the three as follows:

1) The white marlin (*Makaira albidus*), common up and down the Atlantic coast, which probably occurs elsewhere but is apparently not found in the Pacific. It is a small, slender fish with a long, slender spear. The lobes of its caudal fin are narrow and wide-spreading.
2) The black marlin (*Makaira nigricans*), represented by races in the Atlantic and across the Pacific as well. It is a rather heavy fish with a more humped back than that of the white marlin, and more frequently blackish in color than other species. Pacific races of this fish have a short, thick spear. (One race of the black marlin, the blue marlin, is said to be common about Cuba and the Bahamas, frequent at Montauk in the summer, and rare on the east coast of Florida.)
3) The striped marlin of the Pacific (*Makaira mitsukurii*); a more compressed fish than the black marlin with a longer, more slender spear, an equal or greater hump in the dorsal outline at the shoulders, and the pale vertical stripes on the body are better developed than in other species.

Since that succinct explanation was given, however, some modifications have been suggested, and some of the names have been changed.

1. Jordan and Evermann's "A Review of the Giant Mackerel-Like Fishes, Tunnies, Spearfishes and Swordfishes" includes the *Makaira* species above, as well as *Makaira grammatica, M. georgii, M. lessonae, M. ensis, M. gracilirostris, M. marlina, M. holeii, M. zelandica, M. ampla,* and *M. herscheli,* none of which are any longer considered valid names.

In his 1985 *Billfishes of the World*, billfish expert Izumi Nakamura identified the five marlin species that follow.[2]

1) Black marlin (*Makaira indica*): Found throughout the tropical and subtropical waters of the Pacific and Indian Oceans. Stray individuals have been found to migrate into the Atlantic by way of the Cape of Good Hope, but the existence of Atlantic breeding stocks is unlikely. The single lateral line, visible in juveniles, but obscured in adults, is formed into simple loops. Maximum length: 4.5 m (14.76 ft.); world's record weight: 1,560 pounds. Color: body dark blue dorsally and silvery white ventrally; usually no blotches or dark stripes on body in adults, although light blue vertical stripes may occur in a few fish.
2) Indo-Pacific blue marlin (*Makaira mazara*): Found primarily in the tropical and subtropical waters of the Pacific and Indian Oceans. The single lateral line, visible in juveniles, but obscured in adults, is formed into simple loops. It is the most tropical billfish species and is frequent in equatorial waters. Maximum length: 4.5 m (14.76 ft.); world's record weight: 1,376 pounds. Color: body blue-black dorsally and silvery white ventrally, with about 15 rows pale, cobalt-colored stripes each consisting of round dots and/or narrow bars.
3) Atlantic blue marlin (*Makaira nigricans*): Distributed mainly in the tropical and temperate waters of the Atlantic. It is the most tropical of all Atlantic billfishes and probably more abundant in the western than in the eastern Atlantic, judging by commercial fisheries in both areas. The lateral line system, a complicated network pattern, is visible in juveniles, but cannot be seen in adults. Maximum length: 3.75 m (12.3 ft.); world's record weight: 1,402 pounds. Color: body blue-black dorsally and silvery white ventrally, with about 15 rows pale, cobalt-colored stripes each consisting of round dots and/or narrow bars.
4) Atlantic white marlin (*Tetrapturus albidus*): Ranges over most of the Atlantic, from North Carolina to France in the

2. The range and physical characteristics are from Nakamura; the world's record weights are from the 2011 IGFA record book.

north, across the Equator to northern Argentina and South Africa in the south. Maximum length: 2.9 m (9.51 ft.); world's record weight: 181 pounds. Color: body blue-black dorsally, silvery white splattered with brown laterally, and silvery white ventrally; usually no blotches or marks on body but sometimes more than 15 rows of obscure, whitish stripes.

5) Striped marlin (*Tetrapturus audax*): Found throughout the tropical, subtropical, and temperate waters of Pacific and Indian Oceans. Maximum length: 3.5 m (11.48 ft.): world's record weight: 494 pounds. Color: body blue-black dorsally and silvery white ventrally, with about 15 rows of cobalt-colored stripes, each consisting of round dots and/or narrow bands.

Just as Zane Grey's "Giant Tahitian Striped Marlin" was misnamed (and not a new species anyway), any fish that did not fit Grey's preconceived notions about what constituted a valid species could become, in his words, "a new kind of swordfish." Fishing out of Vairao on his 1928 Tahitian adventure, Grey hooked a gigantic specimen:

> Then came a bulge on the surface, a sort of swirl, and then a large round spout of water, out of which heaved a bill and a head that might have belonged to a rhinoceros. In fact, the long spear, curved upwards, was decidedly like that of the African beast. . . . The swordfish heaved out slowly, two-thirds of his length. He was a gleaming blue on his broad back, as wide and round as that of a large horse. His dorsal was short and long. His sides were green and white with wide bars of purple. I was speechless, paralyzed. Such a swordfish I had never even dreamed of. It took only a glance to see that he was far larger than Captain Mitchell's 976-pound marlin, captured in New Zealand.

It is at this moment that Grey exclaimed, "That's a new kind of swordfish!" If indeed it had a bill like the horn of a rhinoceros and was green with purple stripes, it might have been a new species. But it was more likely to have been a large specimen of *Makaira mazara* with an upturned bill. Fish that get away are almost always larger than the ones that are landed, so when his rod broke and the great fish escaped, Grey was free to speculate on its size and decided that it was 18 feet long and that it weighed 1,500 pounds. He called it a "long-speared marlin—giant Tahitian striped marlin," and wrote: "Where there was one of these fish there would be more, and always a bigger one." Unfortunately for Grey, he would not see a marlin that even came close to that size until his next visit to Tahiti two years later, and although it was not a bigger one, he caught it, and it weighed only 1,000 pounds.

The blue marlin has vertical stripes, the black marlin does not, but stripes notwithstanding, the big marlins are very much alike. The dorsal fin is high just behind the shoulders—and these fish can certainly be said to have shoulders—and has a long, low base, trailing all the way back to the small second dorsal. The marlin's first dorsal can be laid back, unlike the high dorsal of the swordfish, which is permanently erect. Where the swordfish has a single, wide keel on each side of the base of its crescent-shaped tail, the marlins have two smaller keels on each side. The marlins' sword is stout at the base and round in cross section, tapering to a rapier-like point. The sword of the blue marlins is proportionally longer and narrower than that of the black; the black marlin's sword is thicker and heavier, often with a slight downturn. Marlins have small, file-like teeth, and a lateral line that cannot be seen unless the elongate, densely packed scales or skin are removed. The adult swordfish has no scales, no teeth, and no lateral line, even under the skin, but all marlins have teeth and a lateral line, although the term "line" is rather loosely applied to this wiggly item. As with the swordfish, in all species of marlins, females grow considerably larger than males.

It is tempting to delete the above paragraphs in discussing the various scientific names, common names, identifications, and misidentifications of the various marlin species and go straight to a family tree that was published in 2011, rendering all those early versions more or less obsolete. But that would overlook the controversies that surrounded the early nomenclature, and besides, the latest names and relationships, albeit based on modern genetic analyses, could be superseded in the future. Anyway, according to John Graves:

> Our DNA analysis revealed that the blue marlin is much more closely related to the sailfish than to the black marlin. In fact, the genetic differences between the blue and black marlin, both of the

genus *Makaira*, are greater than those between almost any other pair of billfish species. It's true that blue and black marlin look very much alike, and both species exhibit strong sexual dimorphism, where the males stop growing at around 200 to 250 pounds while females can grow to weigh well over 1,000 pounds. However, despite these similarities, there are several morphological differences between the blue and black marlin that are consistent with the large genetic distance between the two species. Probably the most significant difference is in terms of early life history. While all other billfish have an elongated bill at small sizes (merely inches long), blue marlin do not get a bill until they are more than 3 feet long. As a result of morphological and genetic differences, black marlin were moved into their own genus, *Istiompax*.

Among the largest fishes ever caught, marlins are found throughout the world's offshore temperate oceans, usually in (or over) deep water. When they hatch, newborn marlins bear little resemblance to the adults. They are bug-eyed little creatures, with a sail-like dorsal fin and large, wide pectorals. As they mature, the dorsal changes into the familiar form, the pectorals become sickle-shaped, and the upper jaw begins to assume the spearlike shape that characterizes the genus. Larvae have been measured at a half inch in length, and assuming that nothing interrupts the growth curve—a questionable assumption where tiny fishes are concerned—they will keep growing, perhaps for as long as ten years, until they reach their massive adulthood. (Any billfish that weighs more than 300 pounds is a female.)

Barbara Block of Stanford University specializes in tuna, but she also studies other species. With David Booth and Frank Carey, she outfitted three blue marlins with acoustic tags to measure their swimming speeds over time and distance. Off the Kona coast of the island of Hawaii, the marlins were caught, quickly tagged and released, and then followed for 25–120 hours. Most of the time, they found, marlins are slow swimmers, evidently preferring to loaf near the surface at speeds rarely in excess of five miles per hour. But, she says, "the fact that they can strike a bait trolled at 800 cm/second [17.90 mph] leaves no doubt that they are capable of short bursts of high-speed swimming, but we saw no such movements in 160 hours of speed observations." Slow swimming is a necessity for long-distance migration; one of the longest migrations ever recorded for a single fish was a that of black marlin that was tagged off Baja California and recaptured off New Zealand by a Japanese long-liner, 6,213 miles from the release.

It is not clear if Jordan and Evermann actually saw swordfish feeding ("rising through schools of mackerel, menhaden, and other fishes, striking right and left with their swords, until they have killed a number, which they then proceed to devour") or if they took the commotion at the surface as evidence that a swordfish was feeding below, but there is an undisputed eyewitness account of a black marlin at work. Diving in Sodwana Bay on the north coast of Natal, South Africa, Mark Roxburgh speared an amberjack, which pulled off his spear and took refuge behind him when a 10–12 foot black marlin appeared. The marlin charged Roxburgh, who was able to deflect the bill, and the amberjack and the marlin circled the diver a number of times. "Seconds later," relates Roxburgh, "the amberjack dived off at great speed to the bottom, closely followed by the highly agitated marlin. Within an estimated 5 sec the marlin had reached its prey and impaled it on its bill. The marlin then shook the amberjack free and swallowed it. . . . Of further interest was the head shaking motion needed to release the impaled prey. Although the frequency of bill use has not been established it is evident that marlin use the spear to impale prey." In the same article, van der Elst and Roxburgh support this contention by describing the stomach contents of an 86 kg (189 pound) black marlin that contained a little tuna (*Euthynnus affinus*) "that had clearly been perforated by the bill."

Like their flat-bladed cousin, the roundbills have also been implicated in attacks in vessels—but not as often. In 1940, when Gudger's report was published, the confusion with common names had not been completely resolved, so we have to depend on his use of the scientific names—some of which have been changed since then—to make sure we understand which fish he is talking about. No problem with the broadbill, which he calls the "true swordfish," but he calls the white marlin (*Makaira albidus*) "the spearfish" and confesses that in most of the ship or timber piercings, if it was not the flat bill of a swordfish, we don't actually know the identity of the perpetrator. But only big marlins have big rapiers, so while we might not be able to identify the fish by species, we can come pretty close. "Dr. W.K. Gregory," wrote Gudger, "has just returned with the Michael Lerner Australia-New Zealand Expedition of the American Museum of Natural History, Jan.- May, 1939, and brings three accounts of marlin attacks on fishing launches in New Zealand waters. . . . These accidents indicate that anglers in growing numbers are pursuing the marlin in these waters, and that the marlins are seeming to retaliate." Both the Indo-Pacific blue marlin and the black marlin are found in New Zealand waters, but if all we have to go on is a broken-off bill, we will never know whether it was *Makaira indica* or *M. mazara* that "retaliated." It probably doesn't make any difference, as neither species is known to be more aggressive than the other, but Gudger gives many instances where the hole in the boat was *round*, indicating a marlin attack. (A swordfish puncture is a flattened ellipse.)

Thirty miles west of the Cape Peninsula in January 1961, a marlin weighing 580 pounds was taken by long-liners, and Frank Talbot and J. J. Penrith of the South African Museum in Cape Town were called in to identify it. The black marlin (then known as *Istiompax marlina*) had swallowed a whole 75-pound yellowfin tuna (*Thunnus albacares*), and when the tuna was pulled out and examined, it was found to have been speared twice, "once laterally through the body, and once obliquely from behind the pectoral in towards the head." "It seems definite," write Talbot and Penrith, "that the spear of marlins is sometimes used for spearing large fishes for prey." Indeed, the bill might be the best way of distinguishing black from blue marlins. In a 2007 article in the sport fishing magazine *Marlin*, Peter Wright writes, "Black marlin tend to sport larger heads and massive, club-like bills. On the other hand, blue marlins wield more slender, rapier-like bill."

Whatever its shape, the bill does not seem to be absolutely necessary. Gudger illustrates several marlins with broken beaks, but it is not possible to tell if the fish was caught in that condition or if the bill was broken in the process of bringing it in. But in a 1951 note in the ichthyological journal *Copeia*, James Morrow described and illustrated a black marlin with no sword at all; it had evidently lost it in some sort of an accident and the lower jaw extended beyond what was left of the upper. "According to the angler who caught it," Morrow observed, "the fish had experienced no difficulty in taking a kahawai [Australian salmon] bait trolled along the surface, had given a good fight and had acted normally in all respects."

Zane Grey's passion for the big marlins—which he called swordfish to the confusion and consternation of those who would track his accomplishments—came right after his passion for true swordfish and tuna. In the chapter he called "The Royal Purple Game of the Sea" in *Tales of Fishes*, Grey describes the "swordfish" he hooked while fishing out of Avalon:

> The swordfish certainly looked a tiger of the sea. He had purple fins, long, graceful sharp; purple stripes on a background of dark, mottled bronze green; mother-of-pearl tint fading into the green; and great opal eyes with dark spots in the center. The colors came out most vividly and exquisitely, the purple blazing, just as the swordfish trembled his last and died. He was nine feet two inches long and weighed one hundred and eighteen pounds.

Grey was also enormously proud to be the first man ever to reel in 1,000-pounder.In *Tales of Tahitian Waters* he described his heroic efforts to land the fish, and the battle to fend off the attacking sharks:

> One big shark had a hold just below the anal fin. How cruel, brutish, ferocious! Peter made a powerful stab at him. The big lance-head went straight through his neck. He gulped and sank. Peter stabbed another underneath, and still another. Jimmy was tearing at sharks with the long-handled gaff, and when he hooked on he was nearly hauled overboard. Charley threshed with his rope; John did valiant work with the boathook, and Bob frightened me with his daring fury as he leaned far over to hack with the cleaver. We keep these huge cleavers aboard to use in case we are attacked by an octopus, which is not a far-fetched fear at all.

Fortunately they were not attacked by an octopus, and they managed to get the mangled marlin to shore and hoisted it up on the tripod so they could look at it and photograph it. Grey's description of the fish is fascinating, not only because he admired it so, but also because of the things he got wrong:

> His color had grown darker, and the bars showed only palely. Still they were there and helped to identify him as one of the striped species. He was bigger than I had ever hoped for. And his body was long and round. This roundness appeared to be an extraordinary feature for a marlin spearfish. His bill was three feet long, not slender and rapier-like, as in the ordinary marlin, or short and bludgeon-like, as in the black marlin. . . . Right there I named this species, Giant Tahitian Striped Marlin. . . . The pectoral fins were large, wide, like wings, and dark in color. . . . His body, for eight feet was as symmetrical and round as that of a good big stallion. According to my deduction it was a male fish. He carried this roundness back to his anal fin, and there further accuracy was impossible because the sharks had eaten away most of the flesh from these fins to his tail. On one side, too, they had torn out enough meat to fill a bushel basket. His tail was the most splendid of all the fish tails I have ever observed. It was a perfect bent bow, slender, curved, dark purple in color, finely ribbed, and expressive of the tremendous speed and strength the fish had exhibited. This tail had a spread of five feet two inches. His length was fourteen feet two inches. His girth was six feet nine inches. And his weight was 1,040 pounds.

The stripes that caused him to call it "one of the striped species," identify it as *Makaira mazara*, the Indo-Pacific blue marlin. (The black marlin, *M. indica*, is stripeless.) Whatever "deduction" Grey used to determine the sex of the fish was wrong because every fish that size is a female. Grey and Hemingway were probably psychologically incapable of conceding such dominant power and size to a *female*, and both authors believed that the biggest and strongest fish were males. In Hemingway's *Islands in the Stream*, the fishing guide Eddy explains: "This damn fish is so strong because it's a he. If it was a she it would have quit long ago. It would have bust its insides or its heart would burst its roe. In this kind of fish the he is the strongest. In lots of other fish it's the she that's the strongest. But not with the broadbill."

Hemingway also had some very peculiar ideas about the biology of the fish he loved and loved to catch. In *Hemingway's Boat*, Paul Hendrickson, quotes a 1933 *Esquire* article:

Hemingway tells his readers it's his hedging belief that almost every known species of marlin, the white, the silver, the striped, the black, maybe even the blue, are only color and sex and age variations of the same fish. The different colors represent different growth stages, not different marlin species. The white is the first stage and the black is always the last stage. The black marlin is always a female even if in its early life it had been male. . . . Black marlin are very old fish, he explains, and you can always tell by the coarseness of their hide and bill, but above all by the way they tire, after the initial struggle which you swear is going to crack your back when they sound.

Zane Grey and Ernest Hemingway shared more than profound misunderstandings about the gender, appearance, and affiliations of their quarry. A dead, bleeding billfish alongside a fishing boat was almost guaranteed to attract sharks, and of his father's ruined marlin in Tahitian waters, Loren Grey writes: "This saga of 83 days trolling, day after day without a strike, and the giant marlin eaten by sharks, is reputed to have been a partial inspiration for Ernest Hemingway's great Nobel-prize winning novel, *The Old Man and The Sea*." But the half-eaten marlin was more than a fish story for Hemingway. Hendrickson decribes what happened in Bimini waters in 1935:

> They were out on *Pilar* one day in middle May when [Henry] Strater hooked into a trophy black marlin. In forty minutes, he managed to get the fish close to the boat. All hands were trying to get it on board when the first shark or two appeared. That's when Hemingway took out a tommy gun that he'd recently acquired from another Bimini fisherman and started spraying the water—which only had the effect of bringing packs of sharks to *Pilar*'s stern. They came for the blood, which was at first was the blood of one of their own. . . . It took another hour to get the fish in. What got in weighed five hundred pounds. There are a lot of pictures of the fish and most of its lower half isn't there. Whole, it might have weighed twice as much.

In *Fishing with Hemingway and Glassell*, Kip Farrington includes a photograph of Hemingway examining "the 500-pound remains of a huge mutilated blue marlin, which was fought by Henry Strater

Figure 6.3. Henry Strater (*left*) and Ernest Hemingway examine the remains of a blue marlin that Strater reeled in 1936 off Bimini. Hemingway tried in vain to drive off the ravenous sharks by shooting at them with a tommy gun. Photo courtesy of John F. Kennedy Presidential Library.

at Bimini in 1935." He also includes a full-page photo with this caption: "In 1936, when he wasn't fishing for marlin, Ernest Hemingway practiced with his beloved Thompson submachine gun on the pier in Bimini Harbor."

Big-game fishing in New Zealand enjoyed its legendary heyday in the 1920s after Zane Grey visited there on his 190-foot schooner *The Fisherman*. Grey and his relatives fished New Zealand waters for marlin, broadbill swordfish, kingfish, and sharks, but except for the mako and thresher, they hated sharks, and if they ever caught one, they chopped it up. (When a shark takes one their baits, Grey's boatman gaffs it, leans over with his knife, and "after a few powerful slashes, that shark was ribbons.") Grey's exploits recorded by cine-camera were shown throughout the world. In 1936, Grey made—and starred in—a movie he called *White Death*, which was ostensibly about the white shark, but the shark doesn't appear until the end. In this movie, shot in New Zealand and Australia, a gent from an unspecified location on the Great Barrier Reef offers Grey the opportunity to catch a 20-foot white shark that has been terrorizing the local aborigines and has gobbled up several of them, along with the wife and son of the local missionary. Grey agrees to catch the shark, partly to win a bet with a rival fisherman and partly to avenge the death of the missionary, who has also fallen victim to the shark. He hooks the shark, and, as in *Jaws*, to which this film bears no other resemblance, his boat is towed all over the place until the shark tires and can be gaffed. The film ends as the shark—a wooden model painted white—is towed onto the beach, and the romantic leads, neither of whom is Zane Grey, gaze into the sunset.

STRIPED MARLIN

Found throughout Indo-Pacific waters, including the southern tip of Africa, the striped marlin (*Tetrapturus audax*) might very well be the most beautiful of the billfishes. Two of the three larger marlin species are striped, and in some cases, the stripes are baby blue or lavender, but the striped marlin is dark blue with light stripes that do not fade as the fish dies, and they mark *T. audax*—the specific name means audacious or bold—as one of the most spectacular fishes that swims. It sports a particularly tall and rakish dorsal fin, giving it an even more audacious look. When excited or hunting, the striped marlin can flash its stripes with neon intensity. The striped marlin is not as big or hefty as its larger cousins—the world's record is a mere 494 pounds—but it is a fighter. The term "greyhounding" is used to describe repeated low-angle jumps from the water; in his book *Billfish*, Charles Mather calls the striped marlin "the most acrobatic and greyhounds more than any other billfish . . . Striped marlin have been known to greyhound as many as many as seventy to one hundred times before being boated."

And—according to one up-close observation—they don't use their bills in the feeding process. Diving with stripers in Magdalena Bay (Baja California), writes Guy Harvey,

> too many marlins to count circled around, but well more than 20 remained at various distances away from the bait ball. They displayed normal swimming posture and color—the dorsal fin half up, anal fin tucked in and stripes muted. . . . The marlin would accelerate in toward the school, and suddenly the stripes would turn on, becoming so wide and vivid they appeared silver. The fish would extend all fins, but hold the pectorals in a low slung position for increased maneuverability. . . . It was plainly obvious that the bill was not being used to slash at the school of bait. The marlin simply overtook the closest baitfish as it crashed through the school, picking off those not swift enough to escape.

Zane Grey's *Tales of the Angler's Eldorado* laid the foundations for game fishing in New Zealand. In the modern era, New Zealand sport fishery, particularly for striped marlin, flourishes because of the moratorium on commercial fishing for marlin in New Zealand waters. In effect since October 1987, the moratorium says that no marlin can be commercially harvested inside New Zealand's 200-mile exclusive economic zone (EEZ), one of the biggest exclusive economic zones in the world. Commercially caught marlin cannot be landed in a New Zealand port, even if taken outside the EEZ, so marlin in New Zealand are exclusively the preserve of the recreational angler. The warm ocean currents bring marlin close in along the New Zealand coast during the summer months, and the season runs from December to May. Striped marlin are the most numerous species encountered and occur from 90 kg (198 lb.) to 200 kg (440 lb.). The New Zealand and world record stands at 224 kg (493 lb.) as of 2012.

In the November 2005 issue of the magazine *Marlin*, Sam Mossman explains why New Zealand provides anglers with the biggest stripers:

> New Zealand marks the southern extent of the striped marlin's water temperature tolerance, and even then show for only about four months of the year when warm currents meander down from tropical waters to the north. In New Zealand, water temperatures of 64 to 68 degrees are standard for striped marlin fishing (although temperatures hit 75 to 77 degrees in high summer), and they have been caught from waters as cold as 59 degrees.

The law of thermal mass dictates that bodies with more mass conserve core heat more efficiently. In the case of fish, the larger the individual, the cooler the water it can stand. New Zealand's geographic situation in the lower latitudes acts as a sort of thermal sorting sieve for marlin size. In a normal year, only the larger fish feel comfortable on the southernmost edge of their range. (Strong El Niño events can bring very warm currents to New Zealand along with large numbers of smaller fish.)

THE SPEARFISHES

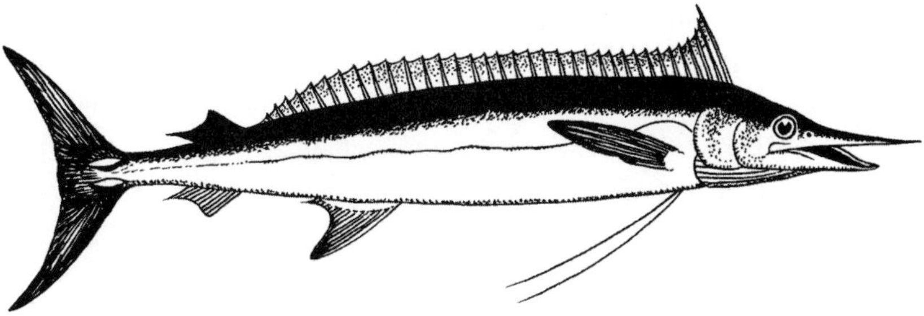

Figure 6.4. Spearfish. Drawing by Richard Ellis.

The striped marlin is *Tetrapturus audax*; the spearfishes have been placed in the same genus. The International Game Fish Association recognizes two species, the longbill spearfish (*Tetrapturus pfluegeri*) and the shortbill (*Tetrapturus angustirostris*) but the *FAO Billfish Species Catalog* by Nakamura adds two more, the Mediterranean spearfish, *T. belone*, and the roundscale spearfish, *T. georgii*. All are about the same size—about six to seven feet long and 70–100 pounds in weight—and they resemble miniature marlins. They are differentiated by slightly different profiles and, in the case of the shortbill, a bill that barely overlaps the lower jaw. This abbreviated bill (again) raises questions of form and function. Such a stubby spear cannot possibly be used to stab prey, and it would be completely ineffective as a club. Why then would such an appendage have evolved? We might assume that the sharp snout functions as a cutwater that enables these slim, graceful creatures to race smoothly through the water, but asking "why" about animal morphology is often a futile exercise. Why, for example, do all other billfishes have pelvic fins while the swordfish has none?

The Mediterranean spearfish, *T. belone*, is the most common billfish in the central Mediterranean and completes its life cycle in this basin. It is characterized by a relatively short bill and a "forehead" that slopes backward from the bill in an almost straight line, as contrasted with the pronounced crest of other spearfish and marlins. Particularly abundant around Italy, this species probably swims in the upper 500 feet of water, generally above or within the thermocline. They travel in pairs, possibly feeding together; their diet consists mostly of fish. This spearfish is probably more widespread in the Mediterranean than was previously believed, as it is easy to confuse this species with the white marlin, *T. albidus*. Known as *aguglia imperiale* in Italian, this species is caught in gillnets, on longlines, by trollers, and by harpooners in the Straits of Messina. It reaches a maximum weight of 100 pounds and is considered a big-game fish. According to a recent report, this species "continues in increasing its importance in the market, in parallel with the increasing of its presence."

Then there is *Tetrapturus pfluegeri*, the longbill spearfish, the smallest of the spearfishes, rarely exceeding 50 pounds in weight. Nakamura describes this species as "remarkably compressed, its depth very low. Bill slender and rather long, round in cross section; nape rather straight. . . ." Identified as a new species only in 1963 (Robins and de Sylva) *T. pfluegeri* is found throughout the North Atlantic south of Newfoundland and Spain, and in the South Atlantic from the mouth of the La Plata to western South Africa. The specific name *pfluegeri* comes from Al Pflueger, Sr., founder of Pflueger Taxidermy in Florida, perhaps the best known of game fish taxidermists. *Angustirostris* means narrow beak (from *angustus*, which is Latin for narrow, and *rostrum*, which, again, means beak, bill, or snout). The shortbill spearfish (*Tetrapturus angustirostris*), as befits its common and scientific names, has the shortest bill of any billfish. It is, however, the most wide-ranging of the spearfishes, found throughout the warm and temperate offshore waters of the Indo-Pacific, from the west coasts of North and South America to Asia and Australia and from the Indian

Ocean to the east coast of Africa. As with the other spearfishes, there is no directed fishery, but many are taken incidentally by long-liners.

WHITE MARLIN

Figure 6.5. White marlin. Painting by Richard Ellis.

The black marlin isn't black, the blue marlin isn't blue, and the white marlin isn't white. (And as we shall see, the white marlin isn't necessarily the white marlin, either.) The black marlin is dark blue above and white below, and the two blue marlins—if there are two distinct species—have a dark blue back with vertical stripes of a lighter blue, sometimes lavender, and white undersides. Why the white marlin should have acquired such a completely inappropriate name is a mystery. (The great white shark, *Carcharodon carcharias*, at a known weight of 2,664 pounds is surely "great," but it is no whiter than a white marlin, except on the undersides.) As for its common name, Migdalski has suggested that "the lateral surfaces lack markings of any kind and appear to be silvery or white, especially when the fish jumps—hence the name."

The white marlin, found only in the waters of the western Atlantic, is the smallest of the true marlins. At a maximum size of about 10 feet and a record weight of 181 pounds, this fish is just a smaller version of its giant relatives, but, to be clear, it is not a juvenile of any other species. Unlike the first dorsals of its larger relatives, that of *Tetrapturus albidus* is rounded at the tip. The white marlin's body is dark blue to chocolate brown, shading to a silvery white underbelly. There are noticeable spots on dorsal fins. *T. albidus* has a visible lateral line that curves above the pectoral fin, then goes in a straight line to the base of the tail. Like every other billfish except the swordfish, the white marlin has paired keels on either side of the tail base.

Fishing for white marlin takes place on both sides of the Atlantic, but it is concentrated on the eastern shore of the United States, from Chesapeake Bay to the Florida Keys, the Caribbean south to northern Brazil. (The world's record white marlin was caught off Vitoria, Brazil, in 1979.) In the eastern North Atlantic, white marlins are sought off the West African island nation of San Tome and Principe, in the Gulf of Guinea, just north of the Equator. Most of the fly-fishing records for *T. albidus* were set by anglers fishing out of the Bom Bom Resort on Principe. By 2007, the white marlin was so heavily overfished that it was considered a candidate for the Endangered Species List.

Now comes the fun part—at least for taxonomists. It seems that some of those overfished white marlins weren't white marlins at all, but roundscale spearfish, previously known from a grand total of four specimens from the eastern North Atlantic and the Mediterranean that had been identified for years as *Tetrapturus albidus*. From the shape of the scales and other almost imperceptible anatomical differences, Mahmood Shivji of Nova Southeastern University's Guy Harvey Research Institute and his colleagues, identified some of the "white marlins" as roundscale spearfish, *Tetrapturus georgii*. Moreover, Shivji, whose specialty is DNA analysis, concluded that the new species is not that closely related to the white marlin—although they look almost exactly alike to fishermen. The roundscale spearfish, however, wasn't exactly unknown. It appears in Nakamura's FAO *Billfishes of the World*, with a drawing that looks unlike a white marlin, but Nakamura writes, "*T. georgei* [*sic*] resembles most closely the white marlin, *T. albidus*, especially in the somewhat humped nape and the broadly rounded anterior lobes of the first dorsal and first anal fins. . . . Further study is strongly needed to clarify the validity of this species."

In a study titled "Effects of Species Misidentification on Population Assessment of Overfished White Marlin and Roundscale Spearfish," Beerkircher et al. state: "Our findings suggest misidentifcations between the species may have affected the accuracy of past *T. albidus* population assessments in the Western North Atlantic, which therefore need re-visiting to permit improved management and recovery of the species." Photographs of a white marlin and a roundscale spearfish side by side (shown here) clearly demonstrate the similarity of the two species and show how easy it is to confuse one with the other. Many of the so-called white marlins caught commercially were probably roundscale spearfish, so the classification of *T. albidus* as potentially endangered needs a serious revisitation.

The fish that had been nicknamed hatchet marlin looked like a white marlin, but where the white had a rounded dorsal fin, hatchets had squared-off dorsal and anal fins that looked as if they had been chopped short. In 2006, John Graves and Jan McDowell of the Virginia Institute of Marine Science examined the DNA of these fish, long considered aberrant white marlins, and concluded that they could be either white marlins or roundscale spearfish. In his article "Billfish Science," Graves writes: "Both white marlin and roundscale spearfish can have truncated dorsal and anal fins, although the cutoff fin shape is more common in roundscale spearfish. In other words, a hatchet marlin can either be a white marlin or a roundscale spearfish. Great!" (The real difference lies in the location of the vent, which is farther forward in the roundscale spearfish than in the white marlin.)

You can now forget everything you thought you knew about the white marlin as *T. albidus* and the striped marlin as *T. audax*. According to the rules of international zoological nomenclature, the white marlin is now officially known as *Kajikia albida* and the striped marlin as *Kajikia audax*. No matter how often they are revised, scientific names are obviously useful in identifying species, because the common names vary from country to country and from language to language, depending on who's fishing for them. The striped marlin, for example, is also known as spearfish, beakie, and barred marlin in English; *empéreux* and *marlin rayé* in French; *agujón, pez aguja, marlin rayado*, and *pez puerco* in Spanish; and *makajiki* in Japanese. No wonder the fishermen are confused.

SAILFISH

Figure 6.6. Sailfish. Painting by Richard Ellis.

In the introduction to his book *The Sailfish: Swashbuckler of the Open Seas*, Jim Bob Tinsley wrote that after he and his wife Dottie caught nine sailfish in four hours of fishing off Stuart, Florida, in 1949, he "wanted to know more about the prized gamester," but discovered that "no complete work on the fish existed, not even an attempt at one." Until this one, Tinsley's book stood alone as the only book devoted to a single species of billfish. Zane Grey came closest with his *Tales of Swordfish and Tuna*, but his book is almost exclusively about fishing for those great game fish, with little or no history or biology. Because he was a fisherman, Tinsley's book has chapters on what you might need in the way of boats, rods, and reels, but it also has sections of early descriptions and depictions of the sailfish, its paleontological history, its breeding and development, and even its natural food and enemies, as well as an 11-page bibliography that explains how he found everything out.

In the essay by George Brown Goode titled "Swordfish" that Zane Grey incorporated into his *Tales of Fishes*, Goode said this about the names of the sailfish:

> The "sailfish," *Histiophorus americanus*, is called by sailors in the South the "boohoo" or "woohoo." This is evidently a corrupted form of "guebum," a name, apparently of Indian origin, given to the same fish in Brazil. It is possible that *Tetrapturus* is also called "boohoo," since the two genera are not sufficiently unlike to impress sailors with their differences. Blecker states that in Sumatra the Malays call the related species, *H. gladius*, by the name "Joohoo" (Juhu), a curious coincidence. The names may have been carried from the Malay Archipelago to South America, or vice versa, by mariners.

Whatever its name in South America or Sumatra, the boohoo, woohoo, or joohoo is now officially known as *Istiophorus albicans* in the Atlantic and *Istiophorus platypterus* in the Pacific and Indian Oceans. (*Istiophorus* means sail-bearer, from the Greek *histion* for sail.) Many believe there is only a single worldwide species (*Istiophorus platypterus*), but if there are two species, they are very similar. (In the FAO *Billfishes of the World*, the same drawing is used to illustrate both species.) The maximum size is 10 feet, and the world's record (Pacific) sailfish, caught off Ecuador in 1947, weighed 221 pounds. (The Atlantic record is 141 pounds.) In *Fishing the Pacific*, Kip Farrington describes the two kinds of sailfishes:

> The bill of the Pacific sailfish is much longer and more tapered than that of the Atlantic variety. The sail is much larger in proportion to

his size and the ventral fin much longer. The coloring is exquisite and no other fish that I have seen, except the Allison tuna, the dolphin and striped marlin, is more beautiful than the Pacific sailfish as it dies. The light blue sheen is indescribable. This is a gallant little fish and I have great admiration for him.

Sailfish are marlin-shaped but slimmer, with a sharp spear that is round in cross section, a blue-backed, silver-bellied color scheme in a pattern of vertical stripes, double keels on the tail stock, and a large, lunate tail fin. But where the marlins have a fairly modest dorsal fin, the sailfishes have the great towering sail that gives them their name, an appendage unlike that of any other fish in the sea. The actual function of the sail is unknown, but we do know it can be laid flat on the fish's back when the fish rockets through the water, and there are records of a sailfish being clocked at 68 miles per hour.[3] Mark Ferrari might be inclined to award the title of "world's fastest fish" to the broadbill swordfish that stabbed him, but certain tunas and the wahoo have also been nominated. It difficult to calcluate the speed of a fish in the water, but even if this could be done, how would we know that it is swimming at its maximum speed? We wouldn't, of course, but Vladimir Walters and Harry Fierstine designed a device that measured the speed of a line as it was taken out by a recently hooked yellowfin tuna and wahoo and found that in the first 10–20 seconds, they were each clocked at around 46–47 mph.

Most authorities believe that the sailfish gets its name from the resemblance of its dorsal fin to a sail (albeit a sort of spiky one, with all those fin rays), but Tinsley cites Sir Thomas Raffles (1881–26), the British East Indian administrator, who claims to have seen the sailfish *sailing*, with its dorsal fin raised to catch the wind. And later, no less a pair of authorities than J. R. Norman and Francis Fraser of the

3. The 68 mph speed usually appears without attribution, but according to Tinsley, "Between 1920 and 1925, Florida's Long Key Fishing Club conducted stop-watch experiments in an effort to determine the speed of the sailfish. The club and its records were destroyed by a hurricane in 1935, but many of the former members recall the experiments and the results. It was found that a hooked sailfish, timed with a stop-watch, could take out 100 yards of loose line in three seconds, a speed of about 68 miles per hour."

British Natural History Museum, in their *Giant Fishes, Whales, and Dolphins,* quote Raffles's description ("The only amusing discovery which we have recently made is that of the sailing fish, called by the natives *ikan layer,* about ten or twelve feet long, which hoists a mainsail, and often sails in the manner of a native boat, and with considerable swiftness"), although they do not verify it. It seems more likely that a sailfish will occasionally lie at the surface with its sail fanned out, sunning itself like a swordfish, and while the sail acts as a heat absorber, it might—by accident, surely—catch the wind. Underwater films have been made of the sailfish with its sail raised to its full height, corralling a school of baitfish, then charging the school and striking at the fish with its sword. The raised sail may also have something to do with mating; Nakamura says that, "around Florida, this species often moves inshore where the females, swimming sluggishly with their dorsal fins extended and accompanied each by one or more males, may spawn near the surface in the warm season."

The spear of the sailfish is smaller and slenderer than that of the marlins, and in no way resembles the flattened weapon of the swordfish, so it is possible (in some cases) to identify the fish that pierces a boat. Gudger's survey contains no specific records of sailfish attacks on boats, but Tinsley's book, devoted as it is to everything sailfish, includes several. The first known attack, he tells us, took place in 1725, when the H.M.S. *Leopard* put in to Portsmouth for repairs, and the narrow bayonet of a sailfish was found in the ship's bottom, having piereced one-inch-thick sheathing, three inches of planking, and another four inches of timber, for a total of eight inches. He quotes the early American big-game fisherman, Charles Frederick Holder, who wrote about sailfish in 1914, as follows: "These magnificent fish are harpooned by the natives of Madagascar and often wreck the boats and kill the men. An American consul saw one leap through the sail of a native proa—and described the fight to me." Tinsley cites several other sailfish attacks on native fishermen and their boats, but it is clear that the sailfish, blessed only with speed and beauty, is no match for the rhino-like charges of the 1,000-pound marlins or swordfishes when it comes to attacking boats.

The capture of the great spearfishes can be dated, often within recent memory. There must have been artisinal fisherman who speared or caught a sailfish from a dugout or outrigger canoe, but in his book, *Salt Water Fishing*, Van Campen Heilner is able to identify with some precision the man who caught the first sailfish for sport. In 1901, Heliner recounts, Mr. and Mrs. Moore were trolling off Soldier Key in Florida, when an unhooked sailfish "jumped to the other side of the boat and in a second or two [another] fish made a jump coming through between the deck stanchions, just where Mrs. Moore had been sitting, struck the iron door of the engine casing, smashing it in, and fell to the bottom of the boat, where he was killed with wrenches and hammers." The fish was seven feet two inches from spear to tail and weighed 87 pounds. Obviously, bludgeoning a sailfish to death with wrenches and hammers is not a particularly good example of "sport fishing," and Heilner tells us that "two or three years before the above time, Mr. Armes, with Mr. Cameron as guide, caught one with rod and reel of about the same size and weight. . . . Therefore we must give credit, so far as we have been able to determine, for catching the first sailfish on rod and reel to Mr. Armes, concerning whom I have been unable to get any further information."

Fishing for sailfish has become somewhat more sophisticated since Mr. Armes's 1898 (or 1899) adventure, and now *Istiophorus* is considered one of the world's premier game fishes—especially for those who would substitute a fly rod for bludgeons. In his *Angler's Guide to the Saltwater Game Fishes*, Ed Migdalski described the appeal of the sailfish:

> A prime favorite with a multitude of fishermen, the sailfish is the billfish easiest to catch and usually the first large fish taken by most budding big game anglers. On the other hand, this fish's leaping, skittering acrobatics after it is hooked are spectacular enough to excite even the most experienced sportsman. . . . It wasn't until the 1920s however, that the Atlantic sailfish was given recognition as one of the gamest fish in existence, and it is now the leading sport fish on the coast of Florida. Today the many undertakings catering to the sportsman in quest of sailfish combine to make it a multimillion dollar industry.

Because they do not achieve the monster size of some of their relatives, sailfishes are popular game for those who do not fancy themselves fighting a ton and a half of fish for half a day or more. The IGFA fly-rod record for Atlantic sailfish is 94 pounds; for Pacific, it's 111 pounds. Philip Wylie, author and fisherman, describes the first sight of a sailfish:

> So he wound [the reel], and looked out on the water, and saw what some men have spent months to witness—the awesome leap of a sailfish, sword flailing, jaws wide, indigo dorsal spread full, the wide-forked tail churning itself free of the water. The fish hung in the air for a second; huge, scintillant, miraculous. Then it was gone. It came again, walking across the sea on its tail.

The popularity of the sailfish among all types of sport fisherman can be seen, for example, in Charles Mather's book *Billfish*, where, alongside a photograph of a young boy with two sailfish, it says: "Small Atlantic sailfish are great sport for small children." The IGFA record book account of the sailfish includes that "its fighting ability and spectacular aerial acrobatics endear the sailfish to the saltwater angler, but it tires quickly and is considered a light tackle species." But because of their brilliant coloration, spectacular acrobatics, and availability, sailfishes are probably the most popular of all big-game fishes. Here Peter Goadby sings the praises of the sailfish:

> They are the peacocks of the sea, with glorious colors and graceful winged bird shape. They are a blend of marine savagery and efficiency in the way they hunt. The beauty and mystery of these acrobatic jumpers—whose muscles can propel their slender needle nose body at a calculated 113 km/h (68 mph)—makes them the light tackle fisherman's dream fish. . . . The dorsal fin sail adds mystique and uniqueness as well as beauty to this fish.

In his authoritative *Fishing Encyclopedia*, A. J. McClane instructs that "most Atlantic sailfish are caught by angling. The favorite method is trolling, using a strip of tuna belly, a whole mullet, or ballyhoo for bait. Sometimes sailfish will take feathers or spoons. A few sailfish may be caught incidental to tuna fishing operations, but the meat tends to be tough, although tasty, and is of little commercial value except when smoked."

As with all other billfishes, larval sailfishes look nothing at all like the sleek sword bearers they will become. In a *National Geographic* article called "Solving the Secrets of the Sailfish," Gilbert Voss states that the juvenile "resembled an adult sailfish as little as a goldfish does a whale shark." At birth, the eighth-inch-long sailfish is a bug-eyed, chubby little creature with a mouthful of tiny teeth and serrated spikes protruding from its gill covers. It feeds initially on minute crustaceans known as copepods, but if it isn't consumed by a larger creature, it begins the process of growing into a swift, sail-backed predator by feeding on small fishes. By three inches in length, the juvenile sailfish has become a miniature replica of an adult, complete with bill and tiny sail. "The number of young sailfish in the oceans would be fantastically large," wrote Voss, "if all the larvae survived, for Marine Laboratory [at the University of Miami] investigations show that a single female may spawn as many as 4,675,000 eggs. However, countless predators feed on eggs and young as they float helplessly in the sea."

7 The Swordfish Mercurial

In a 1979 book called *North Atlantic Seafood,* Alan Davidson identifies the various denizens of the deep that qualify as seafood, including clams, crabs, shrimp, oysters, and lobsters, but the better part of the book is devoted to various fishes and includes a little biology along with the recipes. About the swordfish, he notes that

> Linnaeus compared the flesh of the swordfish to salmon, by way of compliment. The compliment is deserved, but the comparison does not seem entirely apt. The flesh of the swordfish is notably white and firm, and has a fine grain. It is admirably well-suited to being cut into steaks, which may be grilled and served with a thread of lemon juice or good vinegar. They may also be baked, with added liquid. Smoked swordfish is produced in the vicinity of Lisbon. It is served in very thin slices and rivals smoked salmon in quality, although it has a stronger and saltier taste.

But Davidson also says, "There is also the strange affair with the mercury content. It came to light some years ago that swordfish had far more mercury in them than the amount laid down as acceptable for human consumption. Sales in the United States almost stopped as a result. In fact, swordfish had probably been like that for a long time, and it is unlikely that eating it holds any dangers for us." Massachusetts health authorities examined a piece of preindustrial swordfish and found that it contained the same amount of mercury as fishes caught in the 1960s. They also found that the amount of mercury believed to be dangerous to humans was not evident in newly caught swordfish. They therefore ignored the Food and Drug Administration's warnings and allowed swordfish to be landed again. Other states quickly followed, and by 1973, the fishery was fully operational, and the fish, which had been briefly spared, were being caught in unprecedented numbers on longlines. But the fish being brought in were getting smaller and smaller.

Old joke: "What's worse than finding a worm in an apple you're eating?" Answer: "Finding half a worm." Fishermen who prepare their catch for eating and buyers of swordfish in the market often notice whole or partial worms in the white flesh of the fish. (Mysterious holes in swordfish steaks are made by fishmongers removing the worms before selling them.) Fish parasites, found mostly in the internal organs, can be microscopic or plainly visible to the naked eye, and while many species of fish are hosts to various kinds of parasites, worms in swordfish are particularly prevalent and often repulsively visible. Swordfish are hosts to many kinds of parasites including tapeworms (cestodes) in

Figure 7.1. Swordfish prepared for sale in Tokyo's Tsukiji Fishmarket. In Japan, swordfish sashimi is known as *kajiki* or *mekajiki*. Photograph by Richard Ellis.

the stomach and intestines, roundworms (nematodes) in the stomach, flukes (tremodes) on the gills, and copepods attached to the surface of the body. Other parasites include tissue flukes (didymozoidea), gill-worms, (monogenea), and spiny-headed worms (acanthocephalan). Yuck.

For the most part, worms are killed by cooking, but many people are turned off by the thought of eating raw fish (sushi) because there may be live worms or worm eggs in their food.[1] Swordfish is a major component in Japanese fish consumption, but swordfish sushi (*kajiki mekajiki nairagi*) is rare. (There is a restaurant in the Chicago suburb of Batavia named Swordfish Sushi, but even they don't offer raw swordfish.) The high oil content of the white muscle meat of swordfish makes it well-suited for grilling, so most restaurants prefer to serve swordfish steaks.

Are the worms safe to eat? Depends on who you ask. In an article called "Worms in Fish," published by the Florida Fish and Wildlife Conservation Commission, we read, "Although unsightly, these common marine parasites pose no human threat because even if they are not all removed when the fish is cleaned or filleted, they will be killed when the fish is cooked or frozen." But the Seafood Network Information Center of the Sea Grant Extension Program, tells us that

> nematodes rarely cause health problems because they are uncommon in fish fillets and normal cooking easily destroys them. In most cases, swallowing a live nematode is harmless. The nematode passes through the intestine without causing problems. In rare cases, swallowing a live nematode larva can cause severe gastric upset called

1. Not long ago, Americans and Europeans blanched at the thought of eating any kind of raw fish, but nowadays, Japanese sushi and sashimi restaurants are more popular than ever. In cities like New York, Los Angeles, London, and Paris, Japanese cuisine has revolutionized the way westerners think about fish. Theodore Bestor, professor of Japanese studies at Harvard (and the author of *Tsukiji: The Fishmarket at the Center of the World*), points out that sushi has become a global phenomenon, and Japanese food is in vogue in urban Asia as the demand for sashimi and sushi products is on the rise. This has created opportunities for selling sashimi in cities such as Shanghai, Guandong, Beijing, Hong Kong, Taipei, Singapore, Kuala Lumpur, Bangkok, and Ho Chi Minh City.

anisakiasis. This happens when the nematode attaches to or penetrates the intestinal lining. Nematodes do not find humans to be suitable hosts and will not live longer than 7–10 days in human digestive tracts. Swallowing live tapeworm larvae can cause a tapeworm infestation. The tapeworms may live in the human intestinal tract for several years. Symptoms can include abdominal pain, weakness, weight loss and anemia.

You can certainly eat a hunk of swordfish without cooking it, but it really doesn't sound like a good idea. But then there's the mercury problem, which has nothing to with cooking.

Methylmercury—the organic form of mercury found in nearly all seafood—is a potent neurotoxin that can cause nervous system and brain damage in developing fetuses, infants, and young children. Methylmercury (MeHg) also causes neurological damage, cardiac disease, and other birth defects.[2] In his study "The Three Modern Faces of Mercury," Thomas Clarkson of the University of Rochester Medical School tells us that "the first methyl mercury compounds were synthesized in a laboratory in London in the 1860s. Two of the laboratory technicians died of methyl mercury poisoning. This so shocked the chemical community that methyl mercury compounds were given a wide berth for the rest of the century." When the potent antifungal properties of the chemical were discovered, however, it came into wide use, particularly to protect seed grains from devastating fungal infections. The inorganic mercury runoff from treated grains was methylated to methylmercury

2. The mercury scare was initiated by the 1956 discovery that families of fishermen near Minamata on the Japanese island of Kyushu were afflicted with a mysterious neurological disease with symptoms that included loss of coordination, tremors, slurred speech, and numbness in the extremities. The symptoms worsened and led to general paralysis, convulsions, brain damage, and death. The Chisso Corporation chemical factory, which manufactured acetic acid and vinyl chloride and dumped its wastes into Minamata Bay, was identified as the source of the mercury that had contaminated the fish and shellfish. Even when the source of the crippling and often fatal disease was identified, the Japanese government did not order the plant closed until the 1970s. By 1997, more than 17,000 people had applied for compensation from the government. Since 1956, a total of 2,262 people have died of Minamata disease in Japan.

by microorganisms in the sea, thus forging the first deadly link in the marine food chain. Various creatures consume these microorganisms, which are in turn consumed by small fish and cephalopods, then larger predators, and finally the apex predators, such as sharks, dolphins, and the great ocean-ranging fish like tuna and swordfish.

Mercury in humans comes primarily from the consumption of methylmercury-contaminated seafood. A recent study of swordfish in the United States found that about half exceeded the 1 part per million (ppm) safety level set by the Food and Drug Administration (FDA). A study by the Turtle Island Restoration Network also showed that mercury levels in swordfish were significantly higher on average than the FDA safe levels. When Jay Rooker of Texas A&M in Galveston studied the mercury content of Gulf of Mexico fishes caught in proximity to the oil and gas platforms of the gulf, he found that the blue marlin has a far greater concentration of mercury than any other species. (The other species are sharks, cobia, amberjacks, king mackerel, dolphinfish, wahoo, and various tunas.) The marlin gets older and larger than any other inhabitants of the gulf and can therefore accumulate more mercury by eating the larger mercury-consuming prey species. (Swordfish are also found in the Gulf of Mexico, but apparently are not targeted by recreational fishers and, therefore, not submitted to Rooker for analysis.) The mercury in marlin flesh has been tested at 8 ppm, while that of the other large species, such as the bluefin and yellowfin tuna, does not get beyond 1 ppm, the level that the FDA says is safe for human consumption.

Fish and shellfish contain high-quality protein and other essential nutrients, are low in saturated fat, and contain omega-3 fatty acids. Some fish also supply the "good" type of cholesterol, high-density lipoprotein. The "bad" cholesterol (low-density lipoprotein) is believed to be responsible for clogging blood vessels, and high-density lipoprotein is reputed to "exile" the low-density lipoprotein to the liver where the latter is destroyed. In *New Scientist* for August 2002, Meredith Small wrote an article titled "The Happy Fat," in which she identified some of the ways in which the omega-3 fatty acids found in fishes like salmon and menhaden (and flax seeds, walnuts, and olive oil) are now being investigated as an antidote to depression and even as a possible inhibitor of prostate cancer. In a 2002 *Newsweek* article about salmon, Jerry

Adler wrote, "Even people that don't like salmon know by now that it contains omega-3 fatty acids, which are believed to protect against cancer and cardiovascular disease." But some fish and shellfish also contain mercury.

A well-balanced diet that includes a variety of fish and shellfish can contribute to heart health and children's proper growth and development. Women and young children in particular are advised to include fish or shellfish in their diets, but it is precisely women and young children who would be most affected by the mercury in fish and shellfish. Some fish and shellfish contain higher levels of mercury that may harm an unborn baby or young child's developing nervous system. In 2002, Health Canada, a watchdog organization in the Canadian government, warned consumers to limit their intake of certain types of fish because of high mercury levels. The department had previously warned consumers last year about eating shark, swordfish, and fresh and frozen tuna. Certain canned tuna is not affected because younger fish are used in the product and have not accumulated higher levels of mercury in their bodies. The advisory says:

- Mercury is a toxin that can attack the nervous system. It accumulates in the body and can affect fetal development, cause blindness and other birth defects.
- Canadians should limit consumption of fresh and frozen fish to once a week.
- Pregnant women, women of child-bearing age and young children should eat the fish no more than once a month.

The FDA and the Environmental Protection Agency (EPA) are advising some people to avoid some types of fish and to eat fish and shellfish that are lower in mercury. In 2004, the EPA and the FDA issued these guidelines specifically directed toward women who are pregnant or who might become pregnant, nursing mothers, and young children (current levels of mercury in fish and shellfish are said to pose no threat to anyone else):

- Do not eat shark, swordfish, king mackerel, or tilefish because they contain high levels of mercury.
- Eat up to 12 ounces (2 average meals) a week of a variety of

fish and shellfish that are lower in mercury. Five of the most commonly eaten fish that are low in mercury are shrimp, canned light tuna, salmon, pollock, and catfish.
- Another commonly eaten fish, albacore ("white") tuna has more mercury than canned light tuna. So, when choosing your two meals of fish and shellfish, you may eat up to 6 ounces (one average meal) of albacore tuna per week.
- Check local advisories about the safety of fish caught by family and friends in your local lakes, rivers, and coastal areas. If no advice is available, eat up to 6 ounces (one average meal) per week of fish you catch from local waters, but don't consume any other fish during that week.

But there are still those who maintain that mercury in some fish species is unacceptable and that by allowing the public consumption of these fishes, the EPA has caved in to commercial fishing interests. In September 2004, the Sea Turtle Restoration Project showed excessive mercury levels in swordfish purchased at supermarkets in the Sacramento, California, area. Mercury levels in tested fish were measured in excess of 4 ppm, more than 400 percent higher than the FDA's action level. Environmental groups are calling on grocers and restaurants to stop selling swordfish to the public immediately. "Stores are simply ignoring the data that shows swordfish to be highly contaminated with mercury by continuing to sell it, sometimes at significantly discounted prices," says Andy Peri, an analyst for Sea Turtle Restoration Project. "It's deeply disturbing that Sacramento area grocers are simply ignoring the health science and continuing to poison their customers" Both the Sea Turtle Restoration Project and the State of California have made attempts to protect Californians from exposure to mercury in fish by suing Safeway and other California grocers under Proposition 65. Despite Proposition 65 and the pending lawsuits, which requires grocers to post health-warning signs where mercury-contaminated fish is sold, California stores have neglected to post signs adequately in most of their stores.

Some recent studies have shown that consumption of ocean fish species constitutes no threat at all to either children (born or unborn) or adults. A study published in the *Journal of the American Medical As-*

Figure 7.2. Grilled swordfish, mashed potatoes, zucchini, and maybe a soupçon of mercury. Photograph by Richard Ellis.

sociation in 1998, just after the initiation of the Give Swordfish a Break campaign, indicated that methylmercury in fishes posed no threat at all. Philip Davidson of the University of Rochester, the lead author of a study conducted in the Indian Ocean republic of the Seychelles, noted that mercury did no harm to the more than 700 children who were examined, many of whom ate ocean fish several times a week. Indeed, said Thomas Clarkson, one of the authors of the study published in the *Journal of the American Medical Association*, the health benefits from fish outweigh the small risk factor associated with mercury residues in some fish. In another study in the Faroe Islands, children did show signs of mercury poisoning, but they also ate the meat of pilot whales (a kind of large dolphin), which may be a different problem altogether.

In a study published in 2011, Spanish scientists Torres-Escribiano and colleagues analyzed the "bioaccessibility" (the contaminant fraction that solubilizes during gastrointestinal digestion and becomes available for intestinal absorption) of various predatory fishes (swordfish, tope shark, bonito, and tuna) and concluded that "Hg [mercury] bioacessibilty decreased significantly after cooking," suggesting that the mercury content in raw fish is substantially higher than in cooked fish.

Mercury is bad for you, but the health benefits from eating fish may offset the effects of the small amounts of mercury found in some fish. In a follow-up to the *Journal of the American Medical Association* article, Clarkson and J. J. Strain said that "ongoing epidemiological studies of heavy fish consumers in the Seychelles Islands . . . do not reveal adverse effects. To the contrary, the results of some developmental tests that were conducted on prenatally exposed children indicate beneficial outcomes that correlate with mercury levels during pregnancy." The larger the fish, the more mercury it accumulates, so it is probably wise to avoid large species like sharks, tuna, and swordfish. Also whales, but whale consumption isn't a problem in the United States. In Japan, however, where the effects of high doses of methylmercury resulted in Minamata disease, they are still eating the meat of whales, and it's very, very dangerous. In a 2002 article in *New Scientist*, Andy Coghlan wrote,

> Tests on whalemeat on sale in Japan have revealed astonishing levels of mercury. While it has long been known that the animals accumulate heavy metals such as mercury in their tissues, the levels discovered have surprised even the experts. Two of the 26 liver samples examined contained over 1970 micrograms of mercury per gram of liver. That is nearly 5000 times the Japanese government's limit for mercury contamination, 0.4 micrograms per gram. At these concentrations, a 60-kilogram adult eating just 0.15 grams of liver would exceed the weekly mercury intake considered safe by the World Health Organization, say Tetsuya Endo, Koichi Haraguchi and Masakatsu Sakata at the University of Hokkaido, who carried out the research. "Acute intoxication could result from a single ingestion," they warn in a draft paper accepted for publication in *The Science of the Total Environment*."

Unfortunately, it isn't exactly whale meat they're eating. In an article published in 2003, Coghlan notes that "much food labeled as whalemeat in Japan is actually from dolphins and porpoises, and exceeds the government's legal limit for mercury. . . . Even though it is illegal to mislabel meat, DNA analysis of 17 'whale' products purchased in supermarkets by EIA [Environmental Investigation Agency] investigators in 2002 reveals that 12 were from porpoises and dolphins." The average level of mercury was five times the Japanese government's limit of 0.4 ppm, but one sample was 5,000 times over the limit. In Norway, the only other country in the world still actively whaling, the meat of minke whales—at a maximum length of 30 feet, the smallest of the groove-throated whales—is being served in restaurants. In 2003, Norwegian scientists announced that the meat contained dangerous levels of mercury and polychlorinated biphenyl and warned pregnant women and breast-feeding mothers to avoid it.

Tuna are also large-scale predators, at the top of the food chain, and in the course of their predatory activities, they might accumulate a measure of mercury. After the California attorney general tried to get the tuna-canning companies, including Starkist, Chicken of the Sea, and Bumble Bee, to label the cans with warnings to pregnant and nursing women, California superior court judge Robert Dondero ruled that such labels were unnecessary because the mercury levels weren't high enough to pose any danger. In her May 13, 2006, article in the *Los Angeles Times*, Marla Cone quoted Deputy Attorney General Susan Fiering: "The people who will be most hurt are women who don't know about the FDA advisory on the Internet and don't have access to good medical care so they won't know about the danger of mercury in this fish." In July 2006, the FDA announced that imported canned "chunk light tuna" (mostly skipjack) contained nearly 10 times the level of mercury deemed the cut off level for "low mercury" fish. The canned tuna, much of which came from foreign sources, particularly Ecuador, was most likely to be consumed by children and pregnant women, because it is a government-subsidized food item for low-income families. In other words, while one arm of the US government issues stern warnings about mercury, another arm distributes mercury-laden tuna.

In *The End of the Line*, his book about overfishing, Charles Clover includes a list of those fishes that are safe to eat and those that might not. Of the swordfish, he writes that this species is a "slow to mature top predator that plays an important role in the ecosystem. North Atlantic stock is listed as vulnerable by the IUCN [International Union for

Conservation of Nature]. As with sharks, swordfish and some tunas, consumption should be limited because of concern about mercury." So to be safe rather than sorry, don't eat much chunk light tuna, swordfish, whale meat, or dolphin meat. But if the swordfish populations are recovering, you might see that fish reappearing on menus in the near future. Is it safe to eat? For pregnant women, nursing mothers, and young children, no. For the rest of us? Sure, go ahead if you don't mind a little mercury.[3]

3. Breeder's Choice Pet Foods, a company based in Irwindale, California (east of Los Angeles), with brands such as AvoDerm, which specializes in avocado-based food for dogs and cats. "Nutrient-dense and high in crude fiber," says the AvoDerm website, "avocados aren't called the ACE of fruits for nothing—vitamins A, C and E are there in abundance, as is B6. Avocados are also rich in folate, potassium, niacin, essential fatty acids and many other nutrients essential to good skin and coat health as well as good overall health." You can buy avocado-based cat food ("Avo-Derm formulas for cats and kittens are created with California avocados for healthy skin and coat"), but evidently and the same benefits can also be provided by swordfish. AvoDerm now markets "Wild by Nature Swordfish in Swordfish Consommé," which also contains "a balanced Omega-6 to Omega-3 fatty acid ratio." In addition to swordfish, swordfish broth, and avocado oil, canned swordfish for cats also contains sunflower oil, tricalcium phosphate, guar gum, taurine, vitamins and minerals (vitamin E supplement, zinc oxide, thiamine mononitrate, manganese sulfate, vitamin A supplement, menadione sodium bisulfate complex (source of vitamin K activity), riboflavin supplement, pyridoxine hydrochloride, folic acid, vitamin D3 supplement), potassium chloride, and choline chloride.

8 Industrial-Strength Fishing

Swordfish live in all the world's tropical and temperate waters and sometimes in cold waters, as well as in the Mediterranean, the Sea of Marmara, the Black Sea, and the Sea of Azov. Their range can be roughly defined as between 45°N and 45°S. For the most part, they are an offshore species, found at or near the surface or as far down as 3,000 feet. Because people have been catching and eating swordfish for thousands of years, most of our information comes from fishermen. The ancient Greeks and Romans fished for *Xiphias gladius* in the Mediterranean, and there are now fisheries in every ocean, though not necessarily trawled by fishermen whose homeports are on the shore of that particular body of water. Japanese long-liners ply the North and South Atlantic; Spanish ships work the waters of the South Atlantic, the southeastern Pacific, and the Indian Ocean; and swordfish boats flying the flags of Spain, Italy, Morocco, Algeria, Cyprus, Greece, Malta, and Tunisia operate in the Mediterranean, along with the swordfishermen of Sicily. Thousands of tons of swordfish are caught every year and brought to fish markets in almost every country of the world. The firm white flesh is one of the world's favorite seafood items, and therefore, the habits and biology of the swordfish are subjects of intense scrutiny—particularly by fishermen.

As is evident in its name, the International Convention for the Conservation of Atlantic Tunas (ICCAT) was originally convened to assess the status of various tuna species in the Atlantic Ocean. Included were the bluefin tuna (*Thunnus thynnus*), yellowfin tuna (*T. albacares*), albacore (*T. alalunga*), bigeye tuna, (*T. obesus*), and skipjack (*Katsuwonus pelamis*). None of these species were confined to the Atlantic, and even though the acronym was not particularly useful, it was retained. (In *Song for the Blue Ocean*, Carl Safina ridiculed ICCAT as the "International Conspiracy to Catch all Tunas.") The first ICCAT reports were issued in 1973 and addressed only tuna issues, but soon studies of the marlins, bonitos, and swordfish began to appear and, eventually, other large species that were being caught commercially, such as mako and blue sharks. As of 2005, the official mission statement of ICCAT has been:

> The International Commission for the Conservation of Atlantic Tunas is responsible for the conservation of tunas and tuna-like species in the Atlantic Ocean and adjacent seas. The organization was established in 1969, at a Conference of Plenipotentiaries, which prepared and adopted the International Convention for the Conservation of Atlantic Tunas, which include the Convention, which was signed in Rio de Janeiro, Brazil, in 1966. The official languages of ICCAT are English, French and Spanish.
>
> About 30 species are of direct concern to ICCAT: Atlantic bluefin (*Thunnus thynnus thynnus*), skipjack (*Katsuwonus pelamis*), yellowfin (*Thunnus albacares*), albacore (*Thunnus alalunga*) and bigeye tuna (*Thunnus obesus*); swordfish (*Xiphias gladius*); billfishes such as white marlin (*Tetrapturus albidus*), blue marlin (*Makaira nigricans*), sailfish (*Istiophorus albicans*) and spearfish (*Tetrapturus pfluegeri*); mackerels such as spotted Spanish mackerel (*Scomberomorus maculatus*) and king mackerel (*Scomberomorus cavalla*); and, small tunas like black skipjack (*Euthynnus alletteratus*), frigate tuna (*Auxis thazard*), and Atlantic bonito (*Sarda sarda*).

Not all swordfish were caught by commercial fishers. Someone was bound to improve on the time-tested method of locating sunning swordfish by looking for them from the deck or bridge of a fishing boat, and in the 1970s, airplanes were introduced as part of the non-long-line fishery. "But if finding fish from the air seems a high-tech approach—and it is—the irony is that the boats Charlie works with catch the huge fish by throwing spears," writes Safina, describing his experiences with a spotter pilot. Like swordfish, the giant tuna of the

North Atlantic were harpooned at the surface, hauled on deck, and taken to port. Spotter planes were employed in New England and also in the California swordfishery. Some fishermen believed that the planes gave an unfair advantage to those in contact with the spotters, and eventually, the spotter planes were outlawed.

Despite the use of spotter planes for hunting swordfish and tuna, if you Google "spotter+planes+swordfish," you will find a predominance of references to a World War Two British torpedo bomber known as the *Fairey Swordfish*. Developed by the Air Ministry as a torpedo/spotter/reconnaissance plane in 1934, this swordfish was a large biplane with a top speed of 138 mph and a 45-foot wingspan that could be folded for carrier use. Flying from the carrier HMS *Illustrious* in November 1940, Swordfish flew against the Italian navy at Taranto and played an important part in the sinking of the German battleship *Bismarck* in

Figure 8.1. The *Fairey Swordfish*, a British torpedo bomber used by the Fleet Air arm of the Royal Navy during the Second World War. Affectionately known as the Stringbag by its crews, it was outdated by 1939, but archived some spectacular successes during the war, notably the sinking of one battleship and damaging two others of the Italian Navy and the famous crippling of the Bismark. Collection Richard Ellis.

May 1941. As far as I can tell, nobody ever flew a spotter *Swordfish* to spot a swordfish.

In June 1977, a seminar called "Swordfish Workshop" was held at the Southeast Fisheries Center in Miami because of the rapid development of a commercial and recreational fishery for swordfish off the southwest Florida coast. "In the spring of 1976," wrote Grant Beardsley in his summary of the workshop, "several Cuban-American fishermen who had been displaced from the Bahamian lobster fishery, began to experiment with longlines off Miami. Because their success was widely publicized, more and more fishermen entered the fishery." Beardsley continues:

> Previous research in the Atlantic has shown that swordfish may be unusually susceptible to heavy fishing by longline gear as evidenced by declining catch rates and steadily decreasing average sizes in the Canadian longline fishery in the late 1960s. . . . With these thoughts in mind, we felt it prudent and timely to conduct a workshop on swordfish. Our goal was to structure a framework of recommended research activities that would be designed to offset potential management problem areas with the resource and with the fishery itself.

During the mid-seventies off the coast of Miami, Cuban long-liners used floating, amber smudge pots, primarily to keep track of the drift of their relatively short longlines. When they noticed that more swordfish were being caught directly beneath the lighted pots, other fishermen started using small battery-powered Japanese lights. At the same time, chemical light sticks had been developed by the American Cyanamid Company in response to the US government's search for a low heat, nonflammable light that could be used where there was the danger of fire. The light in light sticks is created by the interaction of two chemicals in a glass ampoule inside a plastic tube. When the plastic tube is bent or squeezed, the glass ampoule breaks, allowing the chemicals to interact and create energy, which is emitted as a light source in green, yellow, blue, pink, white, or violet. Light sticks were originally deployed by recreational fishers who caught one fish at a time, but the long-liners quickly adopted the technology, and now the miles-long lines attract their unwary prey with lights as well as baited hooks. Of course, the lights attract everything, not only swordfish, so the longlining bycatch includes a vast assortment of unwanted creatures. In a 2008 study of the bycatch of the Greek swordfish fishery, Peristeraki et al. listed "bluefin tuna, dolphin fish, albacore, black skipjack, spearfish, blue stingrays, sea turtles, ocean sunfish, blackfish, atlantic pomfret, [and among sharks], blues, shortfin makos, threshers, tope and white sharks."

Because swordfish are found in almost all of the world's temperate and tropical offshore waters, it soon became obvious that the Florida fishery was only a small piece in the enormous mosaic that is the commercial swordfishery. There are now fisheries in every ocean, with longliners flying the flags of Spain, Portugal, Italy, Morocco, Algeria, Cyprus, Greece, Malta, Tunisia, Turkey, the United States, Canada, Venezuela, Brazil, Uruguay, Chile, Peru, Ecuador, Mexico, the Philippines, Japan, Taiwan, Australia, New Zealand, Sri Lanka, the Seychelles, Senegal, and South Africa. Throughout the world, only the various tuna species are caught in larger numbers and by more fishermen than swordfish. There have been some cases where swordfish have been considered "incidental" to the tuna fishery and have been discarded. A study of Japanese long-liners in the Atlantic during the year 2000 notes that 15,668 swordfish had been discarded, of which some 35 percent were alive. Beginning in 2000, the Japan fishery management authority instructed its Atlantic long-liners "to discard all their swordfish catch form the beginning of February 2000 to avoid over quota." In 2004, the same authors reported that, in the year 2000, 9,041 swordfsh were discarded; in 2001, 7,853, and in 2002, 4,667.[1] A lot of swordfish are dying so that Japan can fill its insatiable diet for tuna sashimi.

1. These figures, like many others, come from ICCAT reports. Officially, the "Collective Volume of Scientific Papers" is available on ICCAT's website at http://www.iccat.int/en/pubs_CVSP.htm. All the reports from 1973 onward are available online, but this disclaimer accompanies the index:

> These volumes contain the scientific reports of the various groups of the SCRS [Standing Committee on Research and Statistics], as well as papers submitted to, and accepted by, the SCRS. Although some of the conclusions may be provisional, the reports of the meetings contained herein can be considered as official proceedings. Other scientific papers contained in this vol-

In 1978, after the mercury scare, American and Canadian swordfishing resumed, catches increased steadily, and the Spanish fleet of long-liners added substantially to this increase. As of a 1988 ICCAT report, the "harvest levels" for North Atlantic swordfish by long-liners during the years 1983–85 were the highest ever recorded. For each of those years, US and Spanish long-liners took in 19,000–20,000 tons, accounting for more than half of all swordfish caught north of the Azores and east of the Grand Banks. In 1988, Spanish and US long-liners harvested 467,000 swordfish. Moreover, the US and Spanish fleets expanded their operations to the central North Atlantic, the tropical regions of the Caribbean, and the northwest coast of Africa.

Since the early 1980s, the commercial catch of swordfish increased eightfold, but the average weight of fish caught dropped from 115 pounds to 60. Many restaurants refused to put swordfish on their menus in an attempt to discourage fishermen from bringing in the smaller fishes. In 1990, swordfish and marlins were added to the species overseen by ICCAT. The Spanish longline fleet lands more swordfish than any other party to ICCAT. In 1997 Spain had a 45 percent share of the Atlantic swordfish fishery, but of the total of 36,378 tons landed and/or imported into Spain in 1997, almost two-thirds was exported, primarily to Italy, where *pesce spada* is a particularly popular menu item. Although the Spanish catch of swordfish declined slightly in recent years, the Spanish fish-processing industry compensated for the decrease in domestic landings with importation of catch from other countries. Spain is also a major player in the eastern Atlantic and Mediterranean bluefin tuna fishery. Of all bluefin tuna landed and/or imported in Spain in 1997, only one-third was consumed in that country; the remainder constitutes a valuable export, and one that is acknowledged to be a primary factor in recent increases in bluefin tuna catches in the Atlantic and Mediterranean.

The American longline swordfishery expanded to the waters north

ume have not been approved or revised by the SCRS. Thus, the content of each paper is the responsibility of the author(s). Many of these papers deal with on-going research and their conclusions are therefore not yet definitive. In order to ascertain the validity of the data and the conclusions contained in the papers, users are requested to contact the authors.

of Puerto Rico and 1,000 miles farther into the Atlantic east of the Lesser Antilles. By 1987, the world swordfish "harvest" had climbed to 16,500 tons; the US portion of that was 3,636 tons, with a dockside value of $27 million. By 1993, the largest vessels were fishing near the Azores. More than half of the swordfish caught weighed less than 50 pounds. Size at sexual maturity is about 1 m (39 inches) for males and three-quarters of that for females. How old a 1-m swordfish might be is unknown but researchers Kume and Joseph regarded swordfish of less than 130 cm (50 inches) as immature. Since the effort required to catch a large or small swordfish on a longline is the same, it would obviously be beneficial to the fishermen to catch the larger specimens, but there is no way to control what takes the baited hook, whether turtles, dolphins, sharks, tuna, albatrosses, or immature swordfishes.

In "Song for the Swordfish," an *Audubon* magazine article that was published at approximately the same time as his book *Song for the Blue Ocean*, Carl Safina asserts:

> Between 1989 and 1996 East Coast swordfish landings plummeted almost 60 percent (from about 11.3 million to 4.9 million pounds). Earnings plunged from almost $36.5 million to just over 17.6 million. Meanwhile, the number of longline hooks set in the Atlantic increased 70 percent between 1987 and 1995—from 6.5 million to more than 11 million. And that was just from American boats, which take only about a third of the North Atlantic broadbill catch.

These numbers are beyond incomprehensible. They demonstrate an appalling ignorance on the part of consumers—should every fish in the ocean be killed so people can eat swordfish steaks?—and the almost pathological unwillingness of swordfishermen to recognize that they were killing the goose that laid the golden egg. The National Marine Fisheries Service warned in 1997 that "if the North Atlantic stock continues to decline at the same rate as it has since 1978 . . . the commercial fishery may not be viable in approximately ten years." In 1996, the International Union for Conservation of Nature added the North Atlantic swordfish to its *Red List of Threatened Species*, and in 2000, this population was officially classified as endangered.

In March 2011, when the European Union banned catches of swordfish less than 42 inches (110 cm) long in the Mediterranean, the government of Malta protested, claiming that the restrictions would adversely affect the 600 Maltese fishermen licensed to catch swordfish. Reporting on the 2011 ICCAT meeting in the March 2012 issue of *Marlin*, Mike Leech writes, "You would think that great conservation measures would come out of these meetings. Not so much. ICCAT is an association of nations that fish commercially in the Atlantic, Mediterranean, and the Caribbean, and nearly all of its representatives are associated in some way with the commercial fishing industry." He continues:

> Supposedly, the U.S. longlining industry will benefit by retaining their ICCAT quota of North Atlantic swordfish, in spite of the fact that they can't ever seem to fill it. This is not a benefit for swordfish, since the unfilled quota gets transferred to other countries. The more longlining that takes place, the more bycatch of marlin, sailfish, spearfish, sharks, and juvenile swordfish.

Despite ICCAT—or perhaps because of it—illegal or outlaw swordfisheries persist, particularly in the Mediterranean, where *pesce spada* has been pursued for millennia. Out of the southern Italian ports of Monopoli and Manfredonia, tuna long-liners supplement their income by catching baby swordfish, some not more than two feet long, and selling them along the roadside in Sicily. The incredible massacre of the baby swordfish can be seen in a 2011 video, but my attempts to verify the source of the video or locate anyone who would be willing to talk about this abominable procedure have been unsuccessful. (The YouTube video is called *La strage degli spadini*, or "the massacre of the swords," but I warn you, it may be too painful to watch.) Of course, removing fish from the gene pool years before they have had a chance to mature and reproduce adversely affects the population; nothing is more effective than eliminating the potential breeding stock if your goal is population reduction.

Neither harpooning a single fish nor long-lining for hundreds is practicable for small-scale fishers, but there is a simple, old-fashioned technique that is responsible for the death of tens of thousands of swordfish, and millions of other creatures, very few of which ever make it into fisheries statistics. The origins of the drift net are lost in history, but herring fishers of the North Sea were using them in the eleventh century, and by the sixteenth, the Dutch developed large industrial vessels for drift-netting in the open sea and processing the catch on board. (The source of much of the material on drift-netting is Simon Northridge's FAO technical report titled *Driftnet Fisheries and Their Impacts on Non-Target Species: A Worldwide Review.*) Originally woven of rope or twine, the nets are now made of monofilament (or multifilament) nylon fibers and are usually deployed at night, so they will be less visible to the fish. They are supported at or near the surface by a float line and hang more or less vertically in the water, often drifting with the currents, hence their name. They are sometimes called drift gillnets because they catch fish that swim into the net and become trapped behind the gills. The size of the mesh determines what can or cannot pass through, so in that sense drift nets are selective, but except for fishes small enough to pass through the mesh, anything and everything that blunders into them will become entangled.

More than longlines, drift nets trap a variety of creatures, and it was this propensity that led to their eventual ban—too many dolphins were being killed because they were unable to echolocate the filamentous nets. Around 1976, Japan developed the first monofilament nets for deep-sea squid netting, but these nets were banned around the coast of Japan three years later because of the bycatch, which resulted in a vast number of other creatures being trapped in the nets and drowned. In 1983, the Greenpeace ship *Rainbow Warrior* was sent to the Bering Sea to document the kill of marine mammals in Japanese drift nets, but with more than 500 drift-net boats operating in the central Pacific, it was a futile gesture. By 1987, the Japanese squid fleet had expanded to 1,200 boats, each deploying up to 30 miles of nets every night during a seven-month fishing season. Some of the nets were 40 miles long. Japanese and Taiwanese boats drift-netting for albacore kill not only the tuna but tens of thousands of other animals as well.

In 1989, conservationist guerilla Sam La Budde made a film called *Stripmining the Seas*, which showed drift nets being hauled aboard and drowned dolphins being kicked overboard and was seen by many people, including the US Congress. Objections to drift-netting kept piling up; in 1989 the United Nations (UN) adopted a resolution to reduce drift-net fleets and ban drift-netting in the South Pacific. Japan and Taiwan were the major culprits; a 1989 report showed that the Japanese squid fleet alone was responsible for the death of a million blue sharks, 240,000 seabirds, and 22,000 dolphins. Over worldwide objections, the Taiwanese continued to deploy their "curtains of death," and they are still being used by pirate drift-netters. A net that has been cut loose or has broken loose does not shut down; it keeps trapping fishes, dolphins, and sea turtles even when there is no one to haul in the net to harvest or release the prisoners. Known as ghost nets, these gigantic net walls keep on fishing long after the mother ship has departed.

Drift nets are cheap and easy to use because they can be deployed from the small, low-powered vessels that are usually used to supply family or village needs. Small drift nets used by artisanal fishers do not present a problem to swordfish, but some countries have developed large-scale drift-net fisheries that target large pelagic species such as tuna and billfishes, and a directed drift-net fishery is a very big problem indeed for swordfish. A swordfish drift net has a mesh size of 18–24 inches, which allows most smaller fishes to swim right through it, but it traps sharks, sea lions, dolphins, and diving birds. As of Northridge's 1991 report, there was a directed drift-net fishery for swordfish off the California coast that took 1,360 tons of swordfish and 3,000 tons of sharks in 1989. Only the swordfish were destined for fish markets; Americans do not eat shark meat, and six million pounds of sharks were dumped overboard, often after their fins were cut off for Asians to use in shark-fin soup.

Long recognized as a prime location for big-game fishing, the offshore waters of Chile offered local fishermen an ideal arena for a swordfish drift-net fishery. Northridge reported that 4,500 tons of swordfish were taken in 1989 by the 500-boat Chilean drift-net fishery. There were other Taiwanese drift-net fisheries for large species such as bluefin and yellowfin tuna and billfish in the Indian Ocean, and throughout the world's temperate oceans, drift-netters continued to harvest every species of fish (and dolphin and turtle) that was unable to escape through the mesh. Drift nets for tuna and swordfish have been used in the Mediterranean since classical times, and in the 1980s Italy's government encouraged the use of drift nets under the

Figure 8.2. Swordfish drowned in a driftnet; ©2012 Norbert Wu/www.norbertwu.com.

(mistaken) impression that these nets—called *spadare*—were less harmful to other species than longlines. A drift-net fleet that had expanded to more than 700 vessels in 1989 took 6,150 tons of swordfish. Until 1998 the Italian swordfish fishery used drift nets, but while most of these have been replaced by longlines, there are still fishermen using the banned *spadare*. Because many of the drift-net fisheries are illegal, it is sad but safe to assume that many more swordfish died in these fisheries than were officially reported.

Confronted with extensive evidence regarding the significant impact by the drift-net fishery on the resources of the high seas, the United Nations General Assembly (UNGA) unanimously adopted UNGA Resolution 44/225 in December 1989. The resolution called for "immediate action to reduce large scale pelagic driftnet fishing activities in the South Pacific with a cessation by 1 July 1991; immediate cessation of further expansion of large scale pelagic driftnet fishing to other high seas outside the Pacific Ocean; and a moratorium on all large scale pelagic driftnet fishing on the high seas by 30 June 1992 . . . with the understanding the measure will not be imposed . . . should effective conservation and management measures be taken . . . to ensure the conservation of the living marine resources."

A study by Di Natale et al. of the Mediterranean swordfish drift-net fleet was commissioned by the Italian government in 1992 and published by ICCAT in 1995. It showed that the target species represents a large majority of the total catches, but "non-commercial species are also caught, while protected species (cetaceans and marine turtles) are sometimes reported by the fishermen and the observers." In addition to fishes ranging in size from bluefin tuna and albacore to wreckfish, the drift nets also trapped (and drowned) thresher sharks, basking sharks, stingrays, beaked whales, pilot whales, striped dolphins and loggerhead turtles. The authors concluded that "after a deep examination of this matter, it seems necessary to face the whole problem of pelagic drifting gear (driftnets and longlines) for a better assessment of the problem, with the purpose to find, step by step, a reasonable management of the pelagic species in the Mediterranean Sea."

According to a report by Tudela et al., a massive—and illegal—drift-net fishery is conducted by Moroccans in the Alboran Sea, that part of the Mediterranean immediately east of the Straits of Gibraltar. Because swordfish enter this region from the Atlantic from April to June, drift-netting here results in a highly productive swordfishery. The Alboran is regarded as an outstanding area for biodiversity in the Mediterranean, and along with the swordfish, the Moroccan nets also trap large numbers of common and striped dolphins, loggerhead turtles, and blue, thresher, and shortfin mako sharks. Monitoring an illegal fishery is difficult and dangerous; the researchers had to bribe crew members on some boats to provide catch information while remaining under cover. From January to September 2003, four monitored boats (out of a total of some 200) hauled in 2,990 swordfish, 237 dolphins, 46 loggerhead turtles, 498 blue sharks, 542 makos, and 464 threshers. (Obviously, there could be no accurate catch statistics for the major part of the fleet; the fishery is illegal and reports to no one.) Both dolphin species are considered endangered in the Alboran, and while the status of the shark populations is unknown, the species drift-netted by the Moroccan boats are particularly susceptible to incidental fisheries.

For the most part, the nations of the world responded to the UN ban in a positive and supportive fashion. Enforcement remains the critical issue, however. Japan, Taiwan, and South Korea continue to push for exemptions to the moratorium for their drift-net fleets, which, if granted, could undermine the UN moratorium. In addition, pirate drift-net fishing continues, particularly in the Indian Ocean, and several states are seriously concerned about the growing practice of reflagging vessels to avoid enforcement. There are still nations that are willing to ignore the ban on drift nets in pursuit of a quick buck—or yen, peso, ruble, or euro. For example, on May 8, 2000, the *Arctic Wind*, a classic high- seas drift-net boat was apprehended and boarded by a team from the US Coast Guard cutter *Sherman*. The vessel was first seen 600 miles from Adak in the Aleutian Islands, attempting to drift-net salmon and sail back to Asia with its cargo undetected. The *Arctic Wind* sought to escape apprehension by leading the Coast Guard in a convoluted chase across the open seas and did not stop until Captain David Ryan of the Coast Guard received permission to fire on the boat, uncovered his guns, and turned them toward the vessel, ready to fire warning shots. As is typical of illegal high seas drift-net boats, this boat tried to mask its identity. It was flying the flag of Honduras (registration expired) but was Korean owned, and the crew was Russian.

In those parts of the world where the swordfish is fished legally, it is a high-ticket item, although not as high as bluefin tuna. In 2011, a 700-pound bluefin tuna caught off Hokkaido was auctioned off for 32.49 million yen, or nearly $400,000. (That is not a typo: *four hundred thousand dollars*.) Those who would order swordfish with citrus salsa or planked swordfish in an expensive restaurant might consider an alternative luxury: that of saving an endangered species from extinction. The so-called regulatory agencies—ICCAT, the European Union, and the General Fisheries Commission of the Mediterranean—have done little to stem the rising tide of swordfish exploitation.

Since the 1960s, pelagic longlines have been the primary gear used to capture swordfish. The area of US commercial swordfishing has expanded to include the entire US Atlantic coastline, the Grand Banks, the Gulf of Mexico, the Caribbean, and the mid-Atlantic Ocean, which makes it possible to catch swordfish throughout the year. Today's deep-water swordfishers, like those unfortunate members of the *Andrea Gail*'s crew who did not return from their voyage in the "perfect storm" of 1991, set out miles of baited longlines. Sebastian Junger's book *The Perfect Storm* (and the movie that followed in 2000) may indeed have been the first time that many people got any idea of how swordfish were caught. Junger says,

> A hooked swordfish puts a telltale heaviness on the line, and when the hauler feels that, he eases off on the hydraulic lever to keep the hook from tearing out. As soon as the fish is within reach, two men swing gaff hooks into his side and drag him on board. If the fish is alive, one of the gaffers might harpoon him and haul him up on a stouter lint to make sure he doesn't get away. Then the fish just lies there, eyes bulging, mouth working open and shut. If it's a good haul there are sometimes three or four half-dead swordfish sloshing back and forth in the deck wash, bumping into the men as they work. A puncture wound by a swordfish bill means a severe and nearly instantaneous infection. As the fish are brought on board their heads and tails are sawn off, and they're gutted and put on ice in the hold. . . . A longliner might pull up ten or twenty swordfish on a good day, one ton of meat. The most Bob Brown has ever heard of anyone catching was five tons a day for seven days—70,000 pounds of

fish. That was on the *Hannah Boden* in the mid-eighties. The lowest crew member made ten thousand dollars. That's why people fish; that's why they spend ten months a year inside seventy feet of steel plate.

In the 2000 film made from Junger's novel, there are several scenes in which swordfish are brought aboard the fishing boat. As the "fishermen" are actors, they would probably not be able to handle a thrashing swordfish, and it would be difficult to catch that many fish anyway, so the fish in the film, while authentic looking, are all fakes. A company called Edge Innovations of Alameda, California, made the fish, and on their website, they say:

> Edge Innovations created all of the animatronic and synthetic swordfish in the biggest blockbuster of the summer of 2000, *The Perfect Storm*. Not a single real swordfish, dead or alive, was used in the movie. Since the movie was based on a true story, the director, Wolfgang Petersen, wanted everything to be as authentic as possible. That's why he chose Edge to create the swordfish, which were indistinguishable from real swordfish in every respect (they even had realistic blood and guts). Edge also created the menacing mako

Figure 8.3. Although they look remarkably alive (or dead), these swordfish were animatronic models made by Edge Innovations for the 2000 movie *The Perfect Storm*. Photograph courtesy of Edge Innovations, Alameda, California.

shark that came on board and tried to make a snack out of Bobby's (Mark Wahlberg) foot.

The *Hannah Boden* was Linda Greenlaw's boat, but the 70,000-pound haul was before her time. The only woman sword-boat captain, she wrote *The Hungry Ocean* a couple of years after Junger's *Perfect Storm*. The fishermen that Junger wrote about all died, so he had to interview other swordfishermen to learn about their trade, but Greenlaw's stories are based on her own experiences and graphically demonstrate the brutal toughness required for the job. Here, her crew is hauling in the longline after a set:

> The fish circled, swimming under the boat as they often do. Carl held the leader, no longer pulling; he waited. When the fish swam out from under us, Carl pulled in another fathom of leader. A dorsal fin cut the surface; then hell broke loose as the fish slashed wildly with its 3-foot-long sword. The fish's bill and back were lit up in blue and purple, and its sides flashed in silver and pink. With two short jerks, Kenny and Ringo sunk their gaff hooks into the head of the fish and pulled it toward the door in the rail. The fish thrashed, and the water flew. Grabbing a 24-inch steel meat-hook, I reached through the door and placed the hook into one of the fish's eye sockets. Peter came from the stern with a second meat-hook, and placed it in the eye socket with mine. Ringo grabbed the bill to prevent it from slashing as we all pulled together to drag the fish onto the deck. The fish slid through the door easily, and I stood and admired it for a minute. "Nice start, about one fifty," Kenny said as he zipped the bill from the fish with a push and pull of the meat saw.

Here Greenlaw speaks for the swordfishermen:

> I have always been happy to comply with regulations set forth by our country's finest scientists and bureaucrats, and to observe boundaries, believing that laws will insure the future of swordfish and swordfishing. What annoys me are the actions taken by groups such as the head chefs of a number of fine restaurants who boycotted swordfish, taking it off their menus in their Give Swordfish a Break campaign. Give *me* a break! I wonder how these chefs keep themselves abreast of the state of the fishery and how they can be so

> conceited to presume they might know better than the fishermen and scientists who have been working together for years to keep the stocks healthy. In my opinion, little Chef Fancy Pants should work at perfecting his creme brulee and leave fisheries management to those who know more about swordfish than how best to prepare it.

One of the people "so conceited" that they presume to advise fishermen and scientists is Carl Safina, a Ph.D. in ecology, and a fisherman himself. In 1990 he founded the Living Oceans Program at the National Audubon Society, where he served for a decade as vice-president for ocean conservation, and is now president of the Blue Ocean Institute, which he cofounded. He is the author of many scientific papers and the brilliant *Song for the Blue Ocean*, in which he decries the damage being done to the ocean and its inhabitants.

Another way to reduce predatory fish populations is to reduce their prey. In a study that was designed to assess the numbers of dolphins in the Ionian Sea (central Mediterrannean), Milanese researchers Bearzi, Politi, Agazzi, and Azzellino found that along with a 25-fold reduction in numbers of bottlenose (*Tursiops truncatus*) and common (*Dephinus delphis*) dolphins, there was also a significant decrease in the other marine megafauna, particularly tuna and swordfish. They noted that, "in the last decade, there was an obvious decline in epipelagic fish to the point that in 2004, due to the complete lack of sardines, the annual 'festival of the sardine' was celebrated . . . by serving farmed gilthead sea bream." Because dolphins surface to breathe, they can be spotted from boats, and as swordfish also spend time at the surface, they too can be counted by observers. "Billfish encounters" (almost always swordfish) were relatively high in 1997, "but collapsed subsequently." The probable cause for this decline, suggest the authors, "is that the stocks of European anchovy and European pilchard, representing important prey items for common dolphins, tuna and swordfish, are being heavily exploited by purse seiners and catches have been known to decline dramatically in the study areas in recent years." Of course, if their prey is decimated, the predators might move to areas where the prey is still abundant, but the purse seiners are usually a step ahead of the dolphins and tuna and have gotten to the areas of abundance first.

It has long been believed that small, schooling fishes, like anchovies or sardines, are the most numerous fish in the world. In some commercial fisheries, such as the anchoveta fishery off Peru, fishes were caught by the hundreds of tons every year, until—for obvious reasons—the fishery collapsed. It is now believed that some deep-sea species, such as lantern fishes or bristlemouths, are the most populous fish species. But because there is no fishery for these tiny fishes, the overall populations can only be estimated. Consider now, though, the swordfish: The single species, *Xiphias gladius*, has been fished intensively in the Mediterranean since around 1000 BC and elsewhere around the world since the eighteenth century. In their 2000 history of the broadbill swordfisheries of the world, Peter Ward and Sue Elscot list the major areas where commercial swordfishing is carried out and give the annual catch in tons for each area (as of 1998):

Mediterranean Sea	14,670
North Atlantic	12, 175
South Atlantic	13,486
Southeastern Pacific	4,402
North Pacific	8,154
Southwestern Pacific	4,193
Indian Ocean	16,735

The total annual catch is 73,455 tons of swordfish, or 146,900,000 pounds. At an average weight of, say, 40 pounds per fish, that works out to 580,760 fish. A half million swordfish every year! Probably the most startling statement in Ward and Elscot's study is this: "There is no clear evidence of swordfish stocks or their fisheries collapsing from over-fishing." In other words, the swordfish is fecund enough to be able to compensate for the removal of more than half a million adults every year, with no apparent effects on the overall population. Unlike anchovies and sardines, where the weight of an individual fish is calculated in ounces, swordfish are big fish, whose weight is calculated in pounds—sometimes in the hundreds of pounds.

The nations with the highest swordfish catches in the North Atlantic are Spain, the United States, Canada, Portugal, and Japan. The nations that catch the most swordfish in the South Atlantic are Brazil, Japan, Spain, Taiwan, and Uruguay. Mediterranean swordfish are now believed to form a separate stock from the Atlantic stocks, but they are not totally isolated. In 1995, the Mediterranean catch accounted for 9 percent of the world total. In the Indian Ocean, swordfish concentrations occur off the coasts of India, Sri Lanka, Saudi Arabia, and eastern Africa. While the Indian fishery was traditionally largest in the eastern Indian Ocean, it has not been productive in recent years. In 1995, the western Indian Ocean dominated the regional catch, which amounted to 15 percent of the world total catch in 1995. At that time, half the worldwide catch of swordfish occurred in the Pacific. The Pacific swordfish fishery took place in five areas: the northwestern Pacific, off southeastern Australia, off northern New Zealand, the southeastern tropical Pacific, and off Baja California, Mexico. As demand in North America and Europe increases and stricter quotas are set in the Atlantic, scientists expect Pacific swordfish will face more intense fishing pressure in coming years.

9 Big Fish versus Big Squid

In 1981, Ronald Toll and Steven Hess of the University of Miami published the results of their examination of the stomach contents of 65 swordfish caught in the Straits of Florida by commercial long-liners and sport fishermen. "Cephalopods were the most important component," they wrote, "both in numbers and weight. Fish remains were of secondary importance." By far the most common cephalopod was the 10-inch-long shortfin squid (*Illex illecebrosus*), a species found throughout the western North Atlantic, from Florida to Iceland. Toll and Hess list *Ommastrephes*, *Thysanoteuthis*, *Onychoteuthis*, and *Histioteuthis* as genera that are known to aggregate; "heavy swordfish predation upon aggregating or schooling cephalopods is similar to reported predation upon schooling fishes." (One cephalopod species not known to aggregate is *Architeuthis*, but the remains of one giant squid—size unspecified—was found in the stomach of one of the Florida Straits swordfish. Exactly how a swordfish would capture a giant squid is unclear.) In their 1993 study, Spanish researchers Angel Guerra, Fernando Simon, and Angel Gonzalez found that large squids such as *Sthenoteuthis*, *Ommastrephes*, and *Todarodes* were the most important components in the diets of 113 swordfish collected in the North Atlantic between the Azores and Madeira.

Scott and Tibbo noted that swordfish slash laterally with their bills while ascending or descending through a school of prey, and Toll and Hess recorded "numerous decapitated squid, and more frequently, oblique slash marks on mantles, thus supporting the postulated foraging behavior. Furthermore, this concurs with the known horizontal orientation of the pelagic squids listed above." It appears that any smallish marine animal that seeks safety in aggregations is susceptible to the slashing attacks of the hunting swordfish. In *Cephalopod Behaviour*, Roger Hanlon and John Messenger apply the term "shoal" to groups of squids that remain together for social reasons and the word "school" to synchronous behavior (velocity and direction) and polarization (parallel swimming). They assert that "shoaling is a common feature of many squid species and is extremely common amongst fishes, about 10,000 species of which are estimated to shoal as some time in their lives." They list the California market squid (*Loligo opalescens*), the Cape Hope squid (*Loligo reynaudi*), and the northern shortfin squid (*Illex illecebrosus*) as species where shoals "apparently range into the thousands and perhaps even tens of thousands." The authors describe the Caribbean reef squid, *Sepioteuthis sepiodea*, as "clearly one of the most social of cephalopods," but the coral reef habitat of this species means that it is largely immune to the attacks of swordfish. Video tapes of *S. sepiodea* show that they post "sentinels" at the ends of massed lines, specifically for the purpose of prey detection and the transmission of predator information, and this suggests—at least by extrapolation—that other shoaling squid species might not be as vulnerable to the incursive predations of *Xiphias gladius* as less vigilant fish species.

Figure 9.1. The Humboldt squid, *Dosidicus gigas*. Illustration by Richard Ellis.

However they accomplish it, swordfish feed on *Dosidicus*. Also known as jumbo or Humboldt squid, *Dosidicus gigas* does not approach the 60-foot length of the giant squid *Architeuthis*, but it is far more aggressive. According to squid expert Gilbert Voss (who believed *Architeuthis* to be a weak, ineffectual creature), the heavily muscled *Dosidicus* is the true terror of the seas. He noted that the "smaller giant squids of the Humboldt Current . . . attain a length of 8–12 feet and 350 pounds. Their bullet-shaped bodies are heavy and strong, with powerful jet funnels and large fins. Their arms and tentacles are massive and strong and with their beaks they can bite oars and boat hooks in two and eat giant tunas to the bone in minutes." Once believed to be susceptible only to the predations of large sharks and sperm whales, the powerful, fast-moving Humbolt squid also falls prey to the apex of apex predators, the broadbill swordfish.

Cephalopods have many ways of escaping from predators, including "inking," which consists of releasing a cloud of ink into the water that confuses the predator into thinking the escaping squid, cuttlefish, or octopus is somewhere (or something) else. Camouflage may be taken to greater extremes in cephalopods than in any other animals, as they can change color, pattern, or shape (or all three) almost instantaneously to blend into the background, resemble another creature, or assume a shape that might scare off the predator. Some squid species are studded with photophores (light cells) that the can switch on or off at will. Many species can burrow in the sand or take refuge in caves or empty shells. And if these subterfuges fail, the potential prey animal can turn on the afterburners and jet away underwater.

Dosidicus gigas is one of the many squid species known colloquially as "flying squid," so named because they can actually take to the air. They do not actually fly, of course, but, like the "flying" fishes, they are able to build up enough speed underwater to escape the restrictions of their native element and launch themselves through the air. Some observers have noted that a squid in flight ejects water from its siphon, which would liken it to a rocket in the air. Many sailors have been surprised to find squid on their decks in the morning, but there are many instances where this counterproductive behavior has been observed in daylight as well. When Thor Heyerdahl and his companions were sailing the *Kon-Tiki* across the Pacific in 1947, they first thought that squid

Figure 9.2. Argentine spearfisherman Ricardo Mandojana displays his conquest: a six-foot-long Humboldt squid. Photograph courtesy of Ana Maria Mandojana.

had come on deck by climbing aboard the raft, but when they found one on the thatched roof of the deck house, they began to wonder. Finally, the mystery was solved:

> One sunny morning we all saw a glittering shoal of something which shot out of the water and flew through the air like large rain drops, while the sea boiled with pursuing dolphins. At first we took it for a shoal of flying fish, for we had already had three different kinds of these on board. But when they came near, and some of them sailed

> over the raft at a height of four or five feet, one ran straight into Bengt's chest and fell slap on deck. It was a small squid. Our astonishment was great.[1]

If flying squid were not unusual enough, there is another characteristic that is almost as unusual as the flight itself: Whereas flying fish leap out of the water and glide more or less individually, squid sometimes soar *in formation*, leaving and entering the water in unison. Even *Dosidicus*, a hunk of armed muscle that can weigh over 300 pounds, can take to the air. In 1964, off the coast of Chile, D. L. Gilbert shot a movie of a 4-foot-long jumbo squid becoming airborne. Stills from the film are reproduced in Gilbert's 1970 paper (Cole and Gilbert), "Jet Propulsion in Squid." Billfish chasing fish—usually dolphinfish, *Coryphaena*—that are leaping out of the water is a favorite subject for game-fish painters, but I'm not aware of anyone painting a marlin chasing squid out of the water.

In their 1998 study, Mexican researchers Unai Markaida and Oscar Sosa-Nishizaki examined 173 swordfish and found that "cephalopods dominated the stomach contents over the first four [commercial gill-net fishing] trips, 85% by number, 90% by weight, and 96% by frequency of occurrence." (Fishes dominated the last four trips made by the researchers later in the year.) Of the way swordfish actually capture their prey, the authors write:

> The use of the "sword" for disabling or killing prey has been recorded by previous workers. . . . In this study, most prey showed evidence of being slashed or cut. Whole king-of-the-salmon [*Trachipterus altivelis*] as large as 160 cm [5+ ft] were folded in the stomach due

1. Their astonishment would have been lessened if they had read Thomas Beale's 1835 book *A Few Observations on the Natural History of the Sperm Whale*. In this book, Beale writes of squid: "I have myself seen, very frequently while in the north and south Pacific, tens of thousands of these animals dart simultaneously out of the water when pursued by albacore, or dolphins, and propel themselves *head first*, in a horizontal direction for eighty or a hundred yards, assisting their progression, probably, by a rotary or *screwing* motion of their arms and tentacles, and which they have the power of thus moving with singular velocity."

to cuts. Small prey such as *Argonatus* spp. showed no damage and probably had been ingested whole.

Squids were identified in the stomach contents by examination of their beaks. (Malcolm Clarke's 1986 guide to the identification of cephalopod beaks is particularly useful in this endeavor.) The squid species were primarily the purple flying squid (*Sthenoteuthis oualiensis*), which has a mantle length of about 15 inches, and the jumbo squid (*Dosidicus gigas*), which can get big enough to turn the tables on a swordfish. *Dosidicus* could certainly turn the tables on *Xiphias* if the squid attacked in a group.

As McGowan observes, "Anecdotal accounts are given of swordfishes slashing their way through schools of fishes and gathering up their incapacitated victims, but inherent problems make direct observations virtually impossible. Consequently, almost all our knowledge of food capture has to be inferred from the study of stomach contents." Of course, the Humboldt squid caught by swordfish are not eight-footers; it is more likely that the swordfish slash and eat squid in the two-foot range. When Unai Markaida examined the stomach contents of 175 swordfish caught off the west coast of Baja California, he found that the predominant prey species was *Dosidicus*, known to the local fishermen as *calamar gigante*. For the most part, this squid has been considered one of the ultimate predators—strong, smart, and fast, capable of taking down almost anything in range with its powerful parrotlike beak and its muscular arms and tentacles. (When fishermen hook one, others often cannibalize it.)

How dangerous is *Dosidicus*? Depends on who you ask. California underwater photographer Scott Cassell has been diving with *calamares gigantes* off Baja since 1995, and in a 2006 *Outside* magazine article, Tim Zimmerman describes Cassell's very first encounter: "As Cassell tells it, one attacked his camera, which smashed into his face, while another wrapped itself around his head and yanked hard on his right arm, dislocating his shoulder. A third bit into his chest, and as he tried to protect himself he was gang-dragged so quickly from 30 to 70 feet that he didn't have time to equalize properly and his right eardrum ruptured." In the same article, Bill Gilly, an electrophysiologist/teuthologist at Stanford University's Hopkins Marine Lab and an expert on the Humboldt squid, Zimmerman recounts Gilly's noticeably different experience with the same animals: "He jumped into the water wearing nothing but shorts, a t-shirt, and a mask. Within moments he saw a group of five squid ascending from the depths, until they formed a perimeter around him. Then, one by one, they reached out and touched his outstretched hand. Gilly said he felt like he was meeting extraterrestrials coming by to say hello."

In October 2005, Unai Markaida, Joshua Rosenthal, and Bill Gilly tagged 996 Humboldt squid in Mexico within the Gulf of California in the vicinity of Santa Rosalia, and another 997 the following April, off Guaymas. Gilly's efforts are part of a project of the network of researchers known as the Census of Marine Life. The project, known as Tagging of Pacific Pelagics, or TOPP, attempts to learn about the diversity and distribution of marine creatures by tagging large Pacific fauna such as tuna, whales, elephant seals, sea turtles, and now squid. These squid were caught on hand lines by commercial fishermen and tagged as soon as they were brought on deck, often in less than 30 seconds. A reward of $50 was offered for each tag later collected by fishermen, and 160 tags were returned to the biologists, usually within a week or two of tagging. Based on the tag recoveries, Markaida, Rosenthal, and Gilly were able to learn a little about the jumbo squid's migration, its growth rate, and its distribution in the gulf, but there is still a lot we don't know about the squid known as *jibia* by the locals.

When Gilly tagged a half-dozen Humboldt squid in the Gulf of California with pop-off satellite tags, he learned that the squid spent a great deal of time in the depths at low oxygen levels, leading him to question how they could get enough oxygen to function as high-speed predators. In a 2004 interview in the *Los Angeles Times*, Gilly described *Dosidicus* as "the most powerful of the squids with the armament for dealing with big prey: thousands of rings and sucker cups. Yet to see them feeding on little fish in the wild is really interesting. They reach out so delicately and grab them as though they're eating snacks at a cocktail party."

An article in the *Monterey County Weekly* quoted Gilly as saying that "there is an astounding number of Humboldt squid in the ocean . . . 10 million living in one particularly dense area off the coast of Santa Rosalia alone." Ten million five-foot-long squid in one area?

That's either a scuba diver's nightmare or a squid fisherman's wildest dream.[2] If *Dosidicus* releases the typical number of eggs for an oceanic ommastrephid—ranging from half a million to a million—then including the paralarvae and juveniles in the number undermines the whole estimate. Many fish species can also produce a million offspring each, but most of the eggs and larvae are consumed long before they reach adult size, so the idea of five or 10 million *Dosidicus* is meaningless. Before we decide that *Dosidicus gigas* will provide relief for fishers and consumers from declining fish stocks, Gilly said that we "really need to know more about the entire life histories of all the species involved—and how they interact at various times—before we begin to get at causes."

In a study published in 2001 ("Distribution and Concentrations of Jumbo Flying Squid off the Peruvian Coast"), Taipe and his fellow researchers found that the squid the Peruvians call *pota* was present in large numbers all along the coast, in all seasons except for summer, when they tend to disperse. The largest individuals caught by Japanese and Korean jigging vessels measured up to 2 m (6.5 ft.) in length, and weighed more than 300 pounds.[3] Between 1991 and 1999 the industrial fleet caught 576,541 tons of *pota*.

It appears that wherever the range of *Xiphias gladius* and *Dosidi-*

2. When I asked Gilly about this "astounding number," he replied, "We came up with a figure of four or five million squid in an area about twenty-five square miles. It was based on the return the of tags we deployed over a five-day period and the number of squid landed commercially during the same days—and the assumption that our tagged squid were caught with the same probability as untagged squid." Because five million squid is still a hell of a lot of calamari, it is worth noting, Gilly continued, that all the *Dosdicus* in this area were not all mature adults but also paralarvae and juveniles, which "form a vital component of pelagic food webs and serve as essential prey for many (maybe nearly all) pelagic fish at some point in their lives."

3. "Jigging" refers to fishing with a type of lure known as a jig, which consists of multiple hooks molded into a sinker, usually enclosed by a soft body designed to entice fish or squid. On modern squid-jigging boats, multiple lights are used—sometimes underwater—to attract the squid to the boat. The jigs are deployed on rotating drums that bring the captured squid aboard the vessel.

cus gigas overlap, the big squid is the prey of choice for the swordfish. Examining the stomach contents of 48 swordfish caught off Chile in 2003, fisheries biologists Christian Ibáñez, Carlos González, and Luis Cubillos found that, in two of the three areas sampled, *Dosidicus* (known here as *jibia*) represented more than 95 percent of the prey of the *pez espada*. (In the third sample area, the squid *Onychoteuthis banksi*, a foot-long, panoceanic species with hooks on its tentacles, was the preferred prey of the swordfish.) We do not know if *Onychoteuthis* lights up, but we know that *Dosidicus* does, and this propensity might very well give the swordfish the advantage it needs in a battle with the big squid. Both squid and swordfish approach the surface at night, and even though the swordfish can pick out tiny flashes of blue light from lantern fishes and other bioluminescent creatures, a big squid that can flare from red light to white light in an instant may be sending an irresistible signal to a hungry swordfish. Humboldt squid flash when they are excited, so the feeding activities of the swordfish might also attract other squid to this strobe-lit nocturnal banquet table.

It is possible that swordfish could attack the powerful *Dosidicus* while the squid is sleeping—not sleeping as we understand it, but a sort of cephalopod version, where the squid descend into cold, oxygen-depleted waters at around 1,000 feet, and go into a torpid (hypometabolic) state. The squid might do this because they can't stay in warm waters during the day or perhaps their usual prey (lantern fish) also descend into deep, colder waters during the day. They might, however, just be putting distance between themselves and the swordfish. After tagging several swordfish off Cabo San Lucas (Baja California), Carey and Robison noted that "swordfish frequently came up during the day to bask on the surface with the tips of their dorsal and caudal fins out of the water for periods of 15–80 min. This basking behavior at a time of day when we might expect them to be at their greatest depth, may be related to the low oxygen concentration at depth." Swordfish, that is, might be adversely affected by the cold, oxygen-poor waters of the depths and may have to come to the surface to warm in the sun or take advantage of the oxygen-rich upper layers. But there is evidence—such as the stabbing of submersibles—that swordfish can be very active at depths up to 2,000 feet. Teuthologists Ron O'Dor, Bill Gilly, and Brad

Seibel plan to catch *Dosidicus* and tag them in the lab with jet pressure tags in a swim tunnel (the equivalent of a squid treadmill), to "calibrate" the squid for swimming speed and oxygen consumption, and then release them in the wild, to see if the squid might be "sitting ducks" for the swordfish when their oxygen level is down.

10 Benchley and Ellis: Swordfishermen

Shortly after my *Book of Sharks* appeared in 1975, I was asked by ABC-TV's *American Sportsman* if I would like to go shark fishing with *Jaws* author Peter Benchley aboard Frank Mundus's *Cricket II*. The Steven Spielberg movie of *Jaws* (screenplay by Carl Gottlieb and Benchley) had not yet appeared on movie screens around the country, but the novel was scary enough to keep a lot of people out of the ocean. Quick synopsis of the movie/novel plot: A great white shark is operating close to the shore of the fictional town of Amity, Long Island (the film was shot in Martha's Vineyard, Massachusetts), picking off swimmers. Sheriff Martin Brody wants to close the beaches, but Mayor Vaughan wants to keep them open so as not to lose the revenue from summer visitors. Captain Quint, the island's notorious shark fisherman, offers to catch the shark, so with marine biologist Matt Hooper and Sheriff Brody aboard, Quint sets out in his dilapidated boat, *Orca*. Curiously, the novel ends with the death of Quint and Hooper and the survival of Sheriff Brody, but in the film, Quint is gobbled up by the shark just before Brody shoots an oxygen tank that the shark is holding in its jaws, blowing the shark to smithereens.

Frank Mundus had already established himself as a "Monster Fisherman" in Montauk, Long Island, specializing in taking his customers—whom he referred to as "my idiots"—in pursuit of sharks. The usual quarry was blue or mako sharks, but Mundus occasionally caught whites, and in 1964, he hooked, harpooned, and finally shot a white shark that was estimated to weigh 4,000 pounds.

Its gigantic head, triangular teeth bared, hung over the bar of Salivar's, a saloon in Montauk. In the pre-*Jaws* period, Peter Benchley, then an associate editor at *Newsweek*, had gone shark fishing with Captain Mundus, who believed that the character of "Quint" in *Jaws* was based on him. He took every opportunity to behave like a lunatic—which he surely was not—in front of his clients, especially if they brought their cameras. He sported a single pirate's hoop earring (more unusual in 1975 than now), and wore one red sock and one green one so he could differentiate port and starboard. In addition, he kept a loaded rifle handy, to dispatch the dangerous sharks if they gave us any trouble as we hauled them in.

Figure 10.1. Peter Benchley (*left*), Richard Ellis (*center*), and Captain Frank Mundus (the model for Quint in *Jaws*), about to set out on a shark-fishing expedition from Montauk, Long Island, 1975. Collection of Richard Ellis.

With a full TV crew (producer Pat Smith, two cameramen, and a sound man) we boarded the *Cricket II* and set out. It was a brilliant April morning—the kind of day my friend Jack Casey would call a "bluebird day"—and when the Montauk lighthouse had dropped below the horizon, we set the first rods. Benchley, a full-fledged literary lion by then, was to have the first shot, so he strapped himself into the fighting chair and let the line play out. It was only a matter of minutes before he felt a strike, and he began the back-breaking process of reeling in a very heavy and very reluctant fish. Because this was all being filmed, Mundus was supposed to be talking in front of the camera about the shark on the line, and, never at a loss for chatter, he guided Benchley through the process. At this stage, of course, we had no idea what was on the end of the line, but Mundus talked as if it was a shark, and maybe even the kind that Benchley had recently made so famous: "OK, now take it slow. . . . Reel it in as you lean forward, then put your back into it . . . sharks have skin like sandpaper, so if he gets tangled in the line, he'll snip it like a thread . . . great white sharks are dangerous even when they're hooked . . . easy . . . take it easy." While the cameramen filmed the fisherman on the afterdeck, Pat Smith, the producer/director (and famed fishing writer), had climbed up onto the flying bridge so he could see the whole scene that was playing out below. That vantage point put him in a position to see the fish before anyone else did, and when it was close to the surface, he saw that it was iridescent purple in a way that no shark was supposed to be—and had something very long sticking out of its nose. Mundus was still lecturing Benchley on the dangerous shark he was bringing up, but when the fish broke the surface Pat recognized it, and with the cameras running and the microphones recording, he shouted, "That son of a bitch is a swordfish!"

We didn't expect a swordfish, so we simply gaffed the fish and brought it aboard, and the whole "shark fishing" episode had to be repeated. Both Benchley and I later caught blue sharks on camera, but they were not nearly as exciting as that first-strike swordfish. The fish weighed upward of 250 pounds, so when we docked, it was butchered and we took home a trunkful of swordfish steaks, along with the film of our expedition. Even in 1975, large swordfish were becoming scarce because of overfishing. The average weight of swordfish caught since the 1980s has dropped from 115 pounds to 60. Today, many restau-

Figure 10.2. As Captain Mundus talked Benchley through reeling in what was believed to be a shark, he hauled in . . . a swordfish. Photograph by Richard Ellis.

rants are refusing to put swordfish on their menus in an attempt to discourage fishermen from bringing in the smaller fishes. (At Manhattan's Fulton Fish Market (orginally on South Street in downtown Manhattan, but now relocated to Hunt's Point in the Bronx), swordfish that are between 50 and 100 pounds are called dogs; from 25 to 50, pups; and under 25 pounds, rats.) In "Song for the Swordfish," an article in *Audubon*, Carl Safina wrote, "These days, most fishers know swordfish chiefly by their absence, by old-timers' stories and black-and-white photos on the walls of long-established harborside bars. The swordfish, also known as the broadbill, may be the fastest-declining creature in the Atlantic Ocean. . . . U.S. longliners claim that Atlantic swordfish can't recover unless all the countries catching them agree to coordinated measures. But in the 1970s, when concern over mercury levels in swordfish forced U.S. and Canadian longliners to stop fishing, the broadbills recovered within a decade." In 1975, though, we were unaware of these controversies and were delighted with Benchley's serendipitous catch. It was a time of innocence, when few people recognized the threats to the oceans' wildlife, and even fewer recognized that eating swordfish might be dangerous to your health.

11 Are Swordfish Endangered?

Swordfishing began thousands of years ago as a nearshore activity, where fishers harpooned the large fish (mostly mature females) as they basked at the surface. Swordfish used to be more abundant in these inshore areas; the fisheries now take place far offshore. Based on what was considered an almost limitless supply, the fishery has been intensive from its inception, but it was not until the introduction of longlines in the 1960s that the swordfish population was significantly affected. In the late 1950s, in the offshore waters of their home islands, Japanese fishermen began using longlines, which consisted of as many as 6,000 hooks suspended from monofilament lines that could be 40 miles long. They were fishing for tuna, a staple of the sushi menu, and quickly extended their efforts into the eastern Pacific, often within sight of the Mexican coast. Before long, they were in the Atlantic. American fishermen heard of their success in bringing in swordfish, but it was not until 1961, when the Norwegian long-liner *Tampen* put into Gloucester, Massachusetts, with four tons of swordfish in the hold, that Americans and Canadians took notice and, shortly thereafter, followed suit. The first Canadian swordfish long-liner was the *George and Pauline*, which took 16 swordfish, and by the next year, 10 more Canadian boats were outfitted with Norwegian gear. One of the Canadian boats took 184 fish in 1961, and the converted American dragger *Gulf Stream* took 35 fish in that first season. It appeared as if the swordfish population of the western North Atlantic was an enormous resource just waiting to be tapped. As Charles Gibson writes in his "History of the Swordfishery of the Northwest Atlantic," "All indications seemed at the time to be that, given seaworthy vessels and tough crews, the swordfishery could become a year-round business."

During the season of 1963, 100,000 pounds (50 tons) of swordfish came into the Fulton Fish Market in New York City, and the Canadians were on record as bringing two million pounds into the United States in the month of July alone, three times the previous year's total. In September, the Canadian vessel *Beiner* unloaded 293 swordfish at a Newfoundland dock, with the dressed weight of 58,662 pounds. Several other Canadian vessels brought in more than 200 fish apiece. Assuming that the "resource" was limitless, American and Canadian swordfishers were perfecting their techniques and their gear, but as they looked forward to future bonanzas, the uncooperative swordfish population began to thin out. Gibson sums up the drop: "From 1964 to 1968, the swordfishery gradually declined. Foreign imports severely impacted the price structure, and the resource, both in tonnage landed and the size of fish taken, shrank. Fish under 35 pounds were now becoming common and were the reason for the coining of a new market term 'rat,' the old term 'pup' having been reserved for fish above 35 pounds but less than 75 pounds." North Atlantic swordfish and swordfishermen were already in decline by 1969, when the US Food and Drug Administration placed a ban on swordfish containing an excess of five-tenths ppm of mercury, which, says Gibson, "killed the fishery from the Mississippi to the Grand Banks."

By the 1980s, long-liners began to store the freshly caught fish on ice and ship the frozen carcasses to distant markets. This practice reached its zenith in the early years of the twenty-first century when individual bluefin tuna were shipped to Japan—often one fish to a plane—and sold for upward of $400,000. Nothing comes close to such a price, paid for red-meat, fatty tuna in Japan, but swordfish sells for higher prices in Japan than in the United States or Europe because it can also be used for expensive sashimi or sushi.

While the harpoon fishery targeted large mature fish, long-liners caught fish (and countless other creatures) indiscriminately, so the swordfish catch soared because swordfish of every size were being hauled in. There is no way to prevent smaller fish from taking the baited hooks; the only restrictions on the fishermen simply prevent them from selling the juveniles commercially. Whether they are sold or discarded, the younger fish die. Swordfish were numerous, but not numerous enough to successfully withstand massive catches of immature fish. Catching fish before they reach sexual maturity means that these fish will never reproduce; removing large numbers of immature individuals is a guaranteed way of drastically reducing the population. As NMFS put it in 1997:

> When the commercial harpoon fishery prevailed, it took a sustainable amount of swordfish year after year . . . only mature fish that had spawned several times were taken, and it was fully compatible with a viable recreational rod and reel fishery. Today, the recreational and harpoon fisheries are gone and the longline fishery is in decline. The reason is that large adult fish capable of spawning are rare.

As the size of the swordfish dropped, so too did the commercial catches. From 1948, when the world swordfish harvest was 7,000 tons, it grew steadily to 37,700 tons in 1970. That year, reports of high mercury levels in swordfish frightened many consumers, causing the demand to drop significantly, and by 1995, the catch was down to 7,276 tons. (In that year, the United States imported 4,681 tons.) Processors in the United States handled 4,549 tons of swordfish meat, valued at $53.4 million. Of this, fillets constituted 2,920 tons and steaks accounted for 1,629 tons. In 1993, Western Europe consumed 35,324 tons of swordfish, primarily in fillets or steaks. While consumption levels are not known for Asia, high-quality swordfish is valued as an ingredient in sashimi or sushi. In 1995, there were more than 1,900 active swordfish vessels in the United States, mostly long-lining vessels, which caught 36,645 tons, or 41 percent of the world total catch of swordfish.

In the North Atlantic, landings of swordfish fell from 5,100 tons in 1988 to 3,150 tons in 1995. The average weight of swordfish caught in 1963 was 266 pounds; by 1995 it was 90 pounds. By this time, it was becoming clear that swordfish populations were in serious trouble. Drift nets had been banned by the UN in 1992—sort of, that is: nets under a mile and a half in length were still legal—but longlines were snagging enormous numbers of swordfish, often females that were immature,

Figure 11.1. Swordfish for sale in the Milan fishmarket.
Photograph by Alessandro De Maddalena.

too small to breed. Moreover, writes Carl Safina, "Longliners in the American Atlantic discard, dead, about 40 percent of the swordfish they catch—the fish are too small to sell. In 1996, Atlantic swordfishers dumped 40,000 dead juvenile swordfish. Like a maladaptive parasite that kills its own host, longline depletion caused the amount of East Coast swordfish brought to plummet almost 60 percent from 1989 to 1996." The American Fisheries Society, a 136-year old organization whose mission is to "improve the conservation and sustainability of fishery resources and aquatic ecosystems by advancing fisheries and aquatic science and promoting the development of fisheries professionals," issued its "Statement on North Atlantic Swordfish," which begins:

> Atlantic swordfish and other highly migratory tunas and tuna-like species are the management responsibility of the International Commission for the Conservation of Atlantic Tunas (ICCAT). Despite the statutory mandate of ICCAT to maintain stocks at their

level of maximum sustainable yield, North Atlantic swordfish have been overfished for at least 10 years, and the population continues to decline. This statement is issued by the Marine Fisheries Section of the American Fisheries Society to express our concern for the condition of this resource and to urge the ICCAT member nations to adopt effective, risk-averse management measures that will allow the stock to rebuild in a reasonable time period.

In the United States, where the problem was first identified, a concentrated campaign to protect the swordfish from overexploitation was quite successful. On January 20, 1998, SeaWeb and the Natural Resources Defense Council, launched Give Swordfish a Break, a campaign to persuade chefs, restaurant owners, cruise lines, and supermarkets to remove swordfish from menus and dining choices.[1] If

1. SeaWeb is a communications-based nonprofit organization that uses social marketing techniques to advance ocean conservation. By raising public awareness, advancing science-based solutions, and mobilizing decision makers around ocean conservation, they are leading voices for a healthy ocean. SeaWeb was founded in 1995, when the Environment Group of the Pew Charitable Trusts agreed on the need to create a new communications-based ocean initiative to bring ocean conservation issues to life for Americans throughout the country. Since 1999, SeaWeb has been an independent 501(c)3 organization that receives funding from a variety of foundations and other sources. Through all of its programs, SeaWeb is motivated by these core beliefs: that the ocean plays a critical role in our everyday life and in the future of our planet; that the ocean affects us all and what we do affects the ocean; and that the ocean's resources are finite and must be preserved to support human life (http://www.seaweb.org). As for the Natural Resources Defense Council (NRDC), its mission statement declares that its "purpose is to safeguard the Earth: its people, its plants and animals and the natural systems on which all life depends. We work to restore the integrity of the elements that sustain life—air, land and water—and to defend endangered natural places. We seek to establish sustainability and good stewardship of the Earth as central ethical imperatives of human society. NRDC affirms the integral place of human beings in the environment. We strive to protect nature in ways that advance the long-term welfare of present and future generations. We work to foster the fundamental right of all people to have a voice in decisions that affect their environment. . . . Ultimately, NRDC strives to help create a new

restaurants wouldn't serve it, wholesalers couldn't sell it, and if the wholesalers wouldn't buy it from fishermen, the fishermen would have nobody to sell to, and would stop catching swordfish. Among those conservation organizations that signed on were the Wildlife Conservation Society, National Audubon Society, National Coalition for Marine Conservation, the Ocean Conservancy, and the World Wildlife Fund. Twenty-seven leading East Coast restaurateurs announced the removal of swordfish from their menus. Leading hotels and cruise lines followed suit, and *Bon Appétit* magazine announced that it would no longer publish swordfish recipes.

Not everybody agreed that swordfish really needed a break, and many fishers continued to catch the fish, and restaurants continued to offer what they bought, but in June 1998, President Clinton called for a ban on the sale and import of North Atlantic swordfish, and by October, NMFS announced a ban on catching swordfish that weighed less than 33 pounds. (This ban resulted in more juvenile swordfish being tossed overboard, but did little to contribute to the recovery of the species.) By November 1999, ICCAT was forced to acknowledge the campaign and determined that swordfish populations should be rebuilt within 10 years and adopted quotas that, if implemented, would go a long way toward restoring the North Atlantic swordfish population. The Give Swordfish a Break campaign officially ended in August 2000 when the US government closed swordfish nursery areas to fishers in US waters.

Still ongoing is Britain's Bite-Back campaign, whose stated goals are to "focus on establishments that sell non-sustainable pelagic fish—shark, marlin and swordfish—and to levy sufficient pressure on supermarket chains to end their trade. Other fish species widely regarded as endangered or threatened, yet which regularly appear on supermarket fish counters and on restaurant menus, will also feature in the programme." A recent posting on their website (www. bite-back.com) applauds their success with swordfish:

---

way of life for humankind, one that can be sustained indefinitely without fouling or depleting the resources that support all life on Earth (http://www.nrdc .org)."

> The UK's largest retailer, Tesco, has chosen to support Bite-Back's recommendations and removed swordfish from its network of stores. . . . Britain's largest fish retailer, Sainsbury's, has removed pre-packed swordfish and marlin from all 540 stores across the country, thanks to an email campaign from Bite-Back, the marine conservation organization. From this week, 335 Sainsbury's stores will de-list pre-packed marlin and 251 stores will become swordfish-free. The news comes at a time when world-wide consumer demand for swordfish and marlin has almost outstripped supply, prompting the fishing industry to chase these increasingly scarce species to the brink of extinction.

Not everybody is interested in limiting the sale of swordfish. From the website of Boston Pride, a New England–based fish distributor comes the example that follows.

> Boston Pride Sashimi Swordfish is caught in the Pacific Ocean and is available year round. The Sword is bled and frozen at sea to a temperature of −40F. This method of handling and freezing produces a superior frozen product with lighter flesh color and smaller more distinct bloodlines. This all translates into a juicier and milder tasting steak. Boston Pride Sashimi Sword Steaks are individually vacuum packed, size graded into 60 oz, 80 oz, and 100 oz steaks with a tolerance of plus or minus 1 oz. and packed into 10 lb cases. They are then graded into 1/2 moon and 1/4 moon steaks which are separated and packed into appropriately marked cases.
>
> Boston Pride Sashimi Sword Steaks are a cut above other frozen Swordfish steaks. They are cut from frozen nape-off Sashimi grade Swordfish fillets into steaks of a uniform one inch thickness, then separated by shape—either half moon or quarter moon. Land frozen steaks are often hand cut from fresh fish and vacuum packed prior to freezing. The result is often misshaped or randomly shaped steaks with heavy bloodlines and dark, stained meat. This product also often has a stronger "fishy" or "bloody" flavor.

In October 2002, ICCAT announced that North Atlantic swordfish had recovered to 94 percent of levels considered healthy over the last three years. This was defined as "94 per cent of the biomass needed to produce maximum sustainable yield (MSY)"—in other words, almost at the level where previously restricted fishing could resume. Of the recovery, an editorial in the *New York Times* for October 13, 2002, commented:

> This is the best news for fish since the striped bass recovery of the late 1980s, and the lesson for future recoveries is much the same: if you leave the fish alone, or at least give them some space, they will repay the effort. . . . The recovery of the swordfish is one of the few bright lights in the otherwise dismal story of overfishing the fishes at the top of the trophic [what the species ate] pyramid.

SeaWeb and Natural Resources Defense Council considered Give Swordfish a Break a great success. In a press release dated October 3, 2002, Vikki Spruill, executive director of SeaWeb, said, "All participants in the Give Swordfish a Break campaign can feel incredibly proud that their actions helped make a difference for this fish. Overfishing and business as usual is not acceptable. This recovery shows that making tough decisions pays off." And Lisa Speer, an NRDC policy analyst said, "This is a real victory for swordfish and shows that we can restore seriously depleted fish." But swordfish are only one strand of the tangled web that characterizes marine ecosystems, and even in the unlikely event that the four-year restaurant boycott actually had such a remarkable effect on North Atlantic swordfish populations, other large fish species were soon shown to be in dire trouble, the extent of which no one expected.

In a 2001 article about historical overfishing, Jeremy Jackson (and 19 other authors) write:

> Ecological extinction caused by overfishing precedes all other pervasive human disturbance to coastal ecosystems, including pollution, degradation of water quality, and anthropogenic climate change. Historical abundances of large consumer species were fantastically large in comparison with recent observations. Paleoecological, archaeological, and historical data show that time lags of decades to centuries occurred between the onset of overfishing and consequent changes in ecological communities, because unfished species of similar trophic level assumed the ecological roles of overfished

> species until they too were or died of epidemic diseases related to overcrowding. . . . Whales, manatees, dugongs, sea cows, monk seals, crocodiles, codfish, jewfish, swordfish, sharks, and rays are other large marine vertebrates that are now functionally or entirely extinct in most coastal ecosystems.

A year after Jackson et al., Villy Christensen and his colleagues from the Fisheries Center of the University of British Columbia analyzed the abundance of various high trophic-level fish (those at the top of the food chain) in the North Atlantic, from 1950 to 1999. They concluded that the biomass of high trophic-level fishes has declined by two-thirds during that 50-year period. In the late 1960s catches increased from 2.4 to 4.7 million tons annually, but by the 1990s, catches had dropped to below two million tons annually. The fishing intensity for high trophic-level fishes tripled during the first half of the time period and remained high during the last half, but as we have seen, fishing out the highest trophic-level species is a recipe for disaster. Christensen and colleagues conclude: "Our results raise serious concern for the future of the North Atlantic as a diverse, healthy ecosystem; we may soon be left with only low trophic-level species in the sea."

"We humans, as 'the new predator on the block,' can take virtually what we like from the sea, and whatever we remove is taken away from other predators," write Daniel Pauly and Jay Maclean in their book, *In a Perfect Ocean*. They continue:

> Unlike us, though, the other predators in the system are unable to turn to other parts of their own food web and certainly cannot turn to rice or potatoes when marine prey are all gone. On our part, over the course of several centuries we have removed nearly all the large whales (with no rebound in most of their North Atlantic populations), and are presently eating into the populations of other top predators, the sharks and tunas, and of fishes lower down the food web such as cod and other ground fish, all much reduced from their initial abundance. . . . In terms of ecosystems, this means that different species will thrive, others diminish or perish, and the relationship among them will all change. It also means that for the top predator—us—there will be less fish of the kind we like to eat.

In May 2003, in the journal *Nature*, Ransom Myers and Boris Worm published an article that shocked the world. Titled "Rapid Worldwide Depletion of Predatory Fish Communities," the article pointed out that the world's large fish species—including tuna, marlins, swordfish, sharks, codfish, and halibut—had been so heavily overfished by industrial fisheries that 90 percent of them are gone. The authors took 10 years to assemble data sets from all the world's major fisheries and constructed trajectories of biomass and composition of large predatory fish communities from four continental shelves and nine oceanic systems, from the beginning of industrial exploitation to the present. To measure the decline in the open ocean, the researchers also gained access to Japanese long-lining data, which represents the world's most widespread long-lining operation, working in every ocean. They wrote:

> Our analysis suggests that the global ocean has lost more than 90% of large predatory fishes. Although it is now widely accepted that single populations can be fished to low levels, this is the first analysis to show general, pronounced declines of entire communities across widely varying ecosystems. Although the overall magnitude of change is evident, there remains uncertainty about trajectories of individual tuna and billfish species.

They concluded that "declines of large predators in coastal regions have extended throughout the global ocean with potentially serious consequences for ecosystems." The Myers and Worm article (with a photo of drift-netted swordfish on the journal's cover) created a groundswell of concern and outrage. The article (and the shocking conclusions) was repeated in virtually every news medium, from radio, television, newspapers, and magazines to the Internet. Andrew Revkin of the *New York Times* wrote, "Even as sought-after species like tuna and swordfish declined, many other less popular fish also experienced enormous drops in numbers as they were caught unintentionally on mileslong lines of baited hooks or in bottom scouring trawls." A month later, *US News and World Report*'s cover story was titled "Empty Oceans: Why the World's Seafood Supply Is Disappearing." Thomas Hayden, author of the article (which is illustrated with my drawings from *The Empty Ocean*), says:

A trip uptown from the fish market to the American Museum of Natural History highlights another change that's hidden from most consumers. Curator Melanie Stiassny, surveying the museum's spectacularly renovated Hall of Ocean Life, apologizes for a diminutive model of a swordfish, shrunk to fit a display of fish diversity. Her "laughably small" replica, just 4 feet long, is less than a third of the length real swordfish can reach, but it's still larger than a quarter of the 2000 Atlantic catch. Swordfish lines don't discriminate between babies and behemoths. But with most of the big fish already gone, we are catching and eating mainly juveniles, keeping them from reproducing and replenishing the stocks. "Some people," Stiassny says, "think that if you're on the Titanic, you might as well get a first-class ticket."

Ransom Myers, who died in 2007, held the Killam Chair of Ocean Studies in the Department of Biology, Dalhousie University, in Halifax, Nova Scotia, and was one of the foremost fisheries scientists in the world. I asked him before his death how swordfish were faring in the light of his 2003 study. He said, "There appears to be an increase in juvenile swordfish, but this has not translated into an increase in adults. They once were very close to the coast here, but they have been gone for 20 years. Where there is only day tuna long-lining, I believe that swordfish increase (we discussed this in our *Nature* paper). Then the Spanish night swordfish long-liners sets come in and take them out. Traditional fisheries guys don't believe this, but it is obvious if you actually look at the data and think." Myers and Worm, in their paper, actually write, "In the open ocean communities, we observed surprisingly consistent and rapid declines, with catch rates falling from 6–12 down to 0.5–2 individuals per 100 hooks during the first 10 years of exploitation. . . . During the global expansion of the longline fisheries in the 1950s to 1960s, high abundances of tuna and billfish were always found at the periphery of the fished area. Most newly fished areas showed very high catch rates, but declined to low levels after a few years."

Most of the swordfish taken recently by the US fishing industry have been immature. In 1999, the US government took measures to protect juvenile North Atlantic swordfish stocks by closing swordfish nursery areas to fishing. Coupled with an international swordfish recovery plan, swordfish populations are on the road to recovery. In July 2005, Ransom Myers had joined with Carl Safina, Andrew Rosenberg. Terence Quinn, and Jeremy Collie in writing "U.S. Ocean Fish Recovery: Staying the Course;" which was published in *Science.* The opening reads: "With many ocean fish populations at unprecedented lows and declining, management should now emphasize population rebuilding."

After a worldwide decline brought about mainly by commercial long-liners, a concern for dwindling stocks brought on government intervention, and closing certain areas to commercial boats enabled the population to stabilize. In some areas, sport fishermen began to catch fish of almost respectable size. In the Miami Swordfish Tournament of 2005, the winning fish weighed 409 pounds. The Give Swordfish a Break campaign ended in 1999 and resulted in a 10-year international rebuilding program and important protection of nursery habitat. One of those protected areas was the Atlantic coast of Florida, which was temporarily closed, even though NMFS recommended that the closure be permanent. ("Six permits for 'experimental' longlining were nearly issued in 2005," said Leech, writing in 2006, "and NMFS recently announced that it would consider issuing such permits in closed areas allowing a limited number of longliners to operate.") The Gulf of Mexico and the Straits of Florida are known spawning grounds for swordfish (and other billfish and tuna), so the closure of these areas to long-liners began to show results within a few years. As the swordfish reappeared, so did the swordfishing tournaments, and the average weight of the fish caught also began to rise. "Swordfishing in south Florida remains alive and well," concludes Leech, but he points out that "recreational swordfishing remains a tiny blip in the overall landing statistics of Atlantic swordfish."

If fish could read the *New York Times*, the remaining broadbills might be happy to learn that their story is "one of the few bright lights in the otherwise dismal story of overfishing," but alas, they are still being extensively long-lined in the Mediterranean and the Pacific, and *pez espada* still appears on menus around the world. Perhaps the swordfish's pugnacity is a function of its dwindling numbers. Maybe all those peculiarities—the sword, the brain warmer, the gigantic blue eyes—have granted this great and enigmatic fish an understanding of its circumstances that has been denied all other fish and most other mammals, including us. Maybe the swordfish is fighting back.

Because it is one of the world's favorite food fishes, the swordfish has been the target of intensive fishing but also of intensive research. The number of popular articles and scientific studies on *Xiphias gladius* is close to overwhelming. Uncountable are the ICCAT papers on swordfish breeding habits; swordfish catch statistics; swordfish age, length, and weight; swordfish bycatch; and the importance of phases of the moon on swordfish. Many of the authors of the swordfish studies are from countries outside the United States, and they are not always in agreement on solutions to the problems—in fact, they are not always in agreement on what the problems are. Within the United States the same confusion exists. In a 2006 article in the environmental magazine *Mother Jones*, Michael Robbins (a former editor of *Audubon* magazine) writes, "What happens when industry insiders write their own regulations? Welcome to the fishing business." (In his only mention of swordfish in the article, Robbins calls it "another victim of longlining," and goes on to say that "swordfish can grow to 1,000 pounds. But by 1995, the average fish landed weighed only 90 pounds, and most were killed before they could spawn. Better management has led to somewhat of a rebound.") The "rebound" has occurred in some areas—the North Atlantic, for example—but not in others.

The swordfish is primarily a warm-water species and, generally speaking, its migrations consist of movements toward temperate or cold waters for feeding in summer and back to warm waters in autumn for spawning and overwintering. Either they migrate to the north and east along the edge of the continental shelf during summer and return to the south and west in autumn, or there are different groups of swordfish migrating from deep waters toward the continental shelf in summer and moving back to deep waters in autumn. On their website, the Canadian Department of Fisheries says only: "Migratory by nature, swordfish are widely distributed throughout the tropical and temperate oceans of the world. They appear in Canadian Atlantic fishing areas in early June and remain until mid-October/November. Offshore fishermen catch them as far west as the Grand Banks." So Canadian Fisheries commissioned a three-year study that began in September 2005:

To tag swordfish to determine their migration patterns in the Atlantic Ocean. Swordfish tagging studies using traditional "spaghetti" tags have been conducted from St. Andrews Biological Station for many years. Results from these studies have been limited as they rely on the fish recapture and information submitted by fishers. While conventional tags tell us the recapture point and how long it took to get there, they tell researchers nothing about the route taken between release and recapture, or the environmental conditions experienced by the fish during its journey. However, the conventional tagging data have indicated that, while the swordfish move north and south through the Atlantic Ocean, there is less movement from west to east than previously thought.

The pop-up satellite tags are basically tiny waterproof computers, programmed to record the depth of swimming, temperature of the water, and daylight length, while enduring dives to more than 3,000 feet. All of this information is relayed to a satellite maintained by Argos (a worldwide system for location and data collection dedicated to studying and protecting the environment) when the tag "pops up" to the surface. The information is then downloaded to a computer where scientists use programs to analyze the data to determine migration patterns of the fish.

Swordfish are a highly migratory species managed under the International Commission for the Conservation of Atlantic Tunas (ICCAT). This commission is made up of 41 countries of which Canada is a member. There are currently 960 licensed swordfish harpoon fishers in the Maritimes Region.

There appears to be a difference of opinion as to whether the world swordfish population is increasing, decreasing, or holding steady. In their 2000 report for the Australian government—which is interested in expanding their commercial fishery for swordfish because it "is a great-tasting fish [that] has excellent storage qualities"—Ward and Elscot write that "there is no clear evidence of swordfish stocks or their fisheries collapsing from overfishing." But in every recent analysis of the detrimental effects of commercial fishing on populations of large predatory fishes, the swordfish is always included. In May 2003, Myers and Worm concluded that "the large predatory fish biomass today is only about 10% of pre-industrial levels"; that is, 90 percent of the world's predatory fish species (cod, tuna, billfishes, swordfish, flatfishes, sharks, skates and rays—fish that people like to eat), have been fished out, and commercial fishers are now working on the remaining 10 percent. Directed fisheries for swordfish and long-liners that target tuna but catch swordfish incidentally must be having an effect on the world populations of *Xiphias gladius*.

In *The End of the Line*, Charles Clover, environment editor of the *Daily Telegraph* (UK), writes,

> Raul Garcia, my contact from WWF Spain and a Galician himself, found the sight of 4-metre (13-ft) swordfish reassuring. What we were seeing, he told me, was a conservation success, though this was likely to be short-lived. ICCAT, which manages catches of swordfish as well as tuna, actually succeeded in pushing through a reduction in swordfish quotas for a few years in the late 1990s, prompted by an alarming slump in North Atlantic swordfish catches and a drop in the size of the fish. . . . Everywhere else in the ocean there are worrying signs. The swordfish is in a worse state in the southern Atlantic. The state of swordfish in the Mediterranean is desperate, with catches based wholly on small fish, many of which have not had the chance to spawn. Records of catches are atrociously kept. The improvement in the north Atlantic shows what can be done. It bucks the trend, but for how long?

There is no question that commercial fishing has brought down the weight of the average swordfish from 400 or 500 pounds in the early twentieth century to 260 pounds in 1960, and 90 pounds in 1996. The numbers of North Atlantic swordfish dropped to less than 50 percent of what they were in 1980, and US commercial fisherman accounted for about a third of the worldwide catch. A major part of the problem was a direct result of overfishing, but there was also a general absence of regulations regarding fish harvest during critical periods, no minimum fish size, and lack of protection for spawning and nursery areas. For the same reasons, the recreational fishery for swordfish in the Atlantic has largely vanished along with the large fish it used to target. For example, only two swordfish were reported caught by recreational fishermen in 1996 from Virginia to Maine.

In November 2002, in the online *National Geographic News* article "North Atlantic Swordfish on Track to Strong Recovery," we read that "North Atlantic swordfish populations, which had been severely depleted by the 1990s as a result of overfishing, have staged a stunning recovery, reports an international regulatory group charged with overseeing their protection." The "international regulatory group" is ICCAT, the International Commission for the Conservation of Atlantic Tuna, which had been given the responsibility of protecting swordfish and marlin as well as tuna. It appears that new (and enforced) regulations on swordfishing have revived a declining population. If indeed a proportion of breeding age females—previously the primary target of sport fishermen—have survived the attack of the long-liners, we might be able to witness the resuscitation of a once-threatened species. Females can produce 30 million eggs, and while a great majority of these eggs do not make it, such fecundity argues for the possibility of an increase in population if the threat of overfishing is reduced.

Tag-return data indicate that swordfish either make limited local movements during the year or return each year to the same feeding grounds. However, some studies have shown that a few swordfish make substantial seasonal migrations. Tagging studies also show daily migration patterns, where swordfish (in the eastern Pacific, anyway) generally stay inshore near the bottom during the day. At dusk, they head seaward, and after sunset, they feed near the surface and return to inshore areas at sunrise. The larger-scale migration patterns of swordfish are still poorly known, and while it is safe to say that swordfish caught in the North Atlantic probably did not come from the North Pacific, it is not clear that the swordfish caught off the California coast did not come from Hawaii—or even farther west. The first archival tag was implanted in a swordfish in Japanese waters:

> An archival tag equipped with sensors for temperature, depth and luminous intensity is an excellent method to elucidate behaviour and migration of marine organisms. . . . The swordfish, which was harpooned with an archival tag encased in a plastic capsule, was released in July 1999 off the east coast of Japan. The fish was recaptured by a harpoon fishing vessel in June 2000, only 103 km from the tagging location and weighed approximately 120-kg. . . .

> The greatest swimming depth was approximated to be 900 m, deduced from ambient water temperature data. The swimming depth and behaviour pattern changed in response to the ambient water temperature.

Everybody agrees that swordfish are "highly migratory," but nobody really understands their migrations, except to say that they are found in colder northern waters during the summer months and year-round in warmer subtropical and tropical waters and that they generally migrate to warmer waters in the winter for spawning. Tuna are schooling fish, and aerial surveys can reveal the movement of large numbers of tuna, but swordfish are loners, and spotting one swordfish at a time will not reveal much about migration patterns. Research into the mass movement of swordfish is just beginning, but to date, the best—indeed, the *only*—way of determining how many swordfish there are in a given area is to count the number that are caught and evaluate that number over time. If fewer fish are caught than were previously caught, there are probably fewer fish left to catch. It appears, however, that we have very little idea of how many are *not* caught.

In fact, after centuries of swordfishing and perhaps 100 years of counting and analyzing the numbers, we really don't have a handle on the number of swordfish swimming in the world's oceans, but as far as we can tell, it is not going down. In the 2012 *Red Data Book* of the International Union for Conservation of Nature, the swordfish is not listed as a threatened species: "Globally, this species has shown a 28% decline over three generation lengths (20 years). The only stock that is not considered to be well-managed is the Mediterranean, which comprises less than 10% of the species' global range. It is therefore listed as Least Concern, as it is below the threshold for a threatened category under Criterion A1."

Throughout recent history, fishermen have shown an almost pathological disregard for rules and regulations that would curtail their fisheries. One need look no further back than the disastrous overfishing of North Atlantic cod, despite repeated warnings from scientists that the codfishermen were fishing their quarry almost out of existence. The codfisheries of New England and Maritime Canada have been completely shut down. No more codfish; no more codfisheries. The Patagonian toothfish (called Chilean sea bass to make it more palatable) is found in the Southern Ocean (also known as the Antarctic Ocean) and "harvested" illegally by almost every fishing nation. As might be expected from a pirate fishery, the population data are poor, but the toothfish is surely being overfished to the brink of endangerment. Remember the orange roughy? Found in dense schools a mile down in the waters south of Africa, South Australia, and New Zealand, *Hoplostethus atlanticus* used to be a popular menu item, until greedy fishers nearly caught all of them. No more orange roughys in the ocean, ergo no more orange roughys on the menu.

For a thousand years, whales were killed for food, oil, and baleen, and when warnings were posted that the whale stocks were becoming drastically diminished, what did the whalers do? They ignored the warnings, argued that the scientists were wrong, and figured out clever ways to continue whaling. (Some whaling nations, like the Soviet Union and Japan, also lied repeatedly about the species and number of whales they actually took; while giving the appearance of adhering to quotas and restrictions, they were actually breaking every rule in the International Whaling Commission book and then falsifying their reports.) Blue whales and Northern right whales are considered endangered; humpbacks and sperm whales are now considered "vulnerable," even though whalers were supposed to have quit killing them half a century ago.

It is true that *Xiphias gladius* was well-known to the ancient Greeks and Romans, but thousands of years later, the modern-day counterparts of natural historians like Aristotle and Pliny are still trying to untangle the swordfish's mysteries. Satellite tags, remotely operated underwater cameras, and DNA analysis are opening new windows into our understanding of these ocean rangers, but we still don't know how many there are, which is the most pressing question. Until we can answer that question, we are well-advised to err on the side of caution; catch fewer swordfish rather than more. That seems self-evident, but history has shown that ignorance has never been much of a deterrent to overexploitation.

At an International Whaling Commission meeting several years ago, when I asked Sylvia Earle (then deputy commissioner of the US delegation, and one of America's foremost advocates for the ocean and its inhabitants) about the unfortunate human inclination to destroy resources that they might be wiser to protect, she asked me to think about timber, coal, fresh water, oil, whales, fur seals, and passenger pigeons. Given (or taking) the opportunity, humans have overexploited just about every natural resource available to them. The great bison herds of the Western plains, once numbering in the tens of millions, were reduced by buffalo hunters to a few hundred and were saved from the oblivion of extinction by a few dedicated conservationists from the New York Zoological Garden (Bronx Zoo.) Other animals rescued from the brink by conservationists are the California condor, whooping crane, Père David's deer, and Przewalski's horse, but it is really too soon to tell if these efforts will last long enough to enable us to remove these animals from the endangered species list.

This kind of ecological awareness may also have saved the swordfish. Long-liners and drift-netters drove *Xiphias* to low enough numbers that a swordfish panic ensued, bringing out the conservationist armies to the rescue of the well-armed fish that once seemed so capable to taking care of itself. Regardless of how well-armed or how beautifully designed, no creature can defend itself against the voracious armada of the commercial fisheries. The beautiful and powerful bluefin tuna, surely one of nature's grandest creations, ends up as rows and rows of frozen carcasses in the Tokyo fish market and, then, as obscenely expensive sashimi in Japanese restaurants. Millions of sharks—almost by definition capable of taking care of themselves—are caught every year and their fins chopped off to make shark's fin soup. Whereas we used to fear sharks, we now fear *for* them.

12 The Swordfish and Global Warming

By now, everybody knows about global warming. There is even a film on the subject—Al Gore's 2006 *An Inconvenient Truth.* From the North Pole to the Antarctic, the world ocean has warmed significantly over the past 40 years. (The North Pole itself, traditionally buried under the Arctic Ocean ice even in summer, has in recent years, been located in open water.) As the waters of the Arctic become warmer and the ice pack diminishes, polar bears, accustomed to swimming from one floe to another, jump in the water to swim to the next floe, but because there *is* no next floe, the bears swim until they become exhausted and drown. As bizarre as it seems, this has become a serious threat to polar bear populations, and we have begun to worry about the survival of the great ice bears. We probably worry more about polar bears than mussels, but along the coasts of central and northern California, the mussel beds (mostly *Mytilus californianus*) that provide a microhabitat and a refuge from predators for hundreds of species, are in serious decline, and with them, the complex biodiversity of the wave-exposed coasts. In 2006, Jayson Smith, Peggy Fong, and Richard Ambrose studied the reduced biodiversity of the mussel beds and correlated it directly to an increase in sea-surface temperature over the past 30 years. Their study is one of the first empirical examples linking the disappearance of species to climate change, but it will probably not be the last.

Because articles on global warming are written by people, the emphasis has almost always been on how rising temperatures will affect the land and those of us who live on it. As melting ice caps cause the oceans to rise, seaports and other sea-level coastal habitations will get swamped; overheating of the earth by greenhouse gases and depletion of the ozone layer will cause massive weather changes, and as the land and the ocean warm up, the balance of ecosystems around the world will be altered. Many kinds of pollution trap heat at the earth's surface like a blanket; as pollution builds, too much heat is being trapped. There will be an increase in disease, air pollution, hurricanes, blizzards, and heat waves. But what will be the effect of global warming on the sea?

The sea is far more absorbent of heat than the atmosphere, and because it covers two-thirds of the planet's surface, the effects of global warming on marine ecosystems are of great significance but poorly studied. "Until recently," write Sydney Levitus et al. in a 2000 study in *Science*, "little work has been done in systematically identifying ocean subsurface temperature variability on basin and global scales, in large part due to the lack of data." After analyzing data from millions of temperature and salinity measurements made around the world during the past century, Levitus and his fellow oceanographers conclude that "a large part of the world has exhibited coherent changes of ocean heat content during the past 50 years, with the world exhibiting a net warming." The oceans are indeed heating up, even though the melting of the polar icecaps introduces large amounts of cool, fresh water into the salty sea. In a study published in 2003, Ruth Curry, Bob Dickson, and Igor Yashayaev measured the salinity at various locations in the Atlantic to ascertain "rates of freshwater fluxes, freshwater transport, and local ocean mixing—important components of climate dynamics." They found a "growing body of evidence indicating that shifts in the oceanic distribution of fresh and saline waters are occurring worldwide in ways that suggest links to global warming and possible changes in the hydrologic cycle of the Earth."

"Is Europe's central-heating system about to break down causing climate chaos around the world?" asks Stephen Battersby in a 2006 *New Scientist* article. "Late last year," he continues, "oceanographers reported a sudden and shocking slowdown in the currents of the North Atlantic, a critical part of the vast system of ocean circulation that influences temperatures and weather around the world." The oceans of the world are all linked by a network of currents sometimes called the global conveyor, with warm surface flow connecting to deep cold currents. An offshoot of the Gulf Stream called the North Atlantic Drift flows all the way to the seas off Greenland and Norway. Evaporation makes the water saltier, so as it is chilled by Arctic winds it becomes denser than the waters underneath it and it sinks. Climate change is injecting more and more fresh water into the Arctic by the calving of icebergs from Greenland, which dilutes the North Atlantic Drift reducing its density and making it more buoyant. Thermohaline circulation depends on heat and salt. If the fresh water input reaches a critical rate and the sinking stops, the northern branch of the conveyor would stop, and warm tropical waters of the Gulf Stream would no longer flow past the west coast of Europe.

"A Worrying Trend of Less Ice, Higher Seas," is the title of Richard Kerr's article in *Science*, dated March 26, 2006. He writes:

> The summertime Arctic Ocean could be ice-free by century's end, 11,000-year-old shelves around Antarctica are breaking up over the course of weeks, and glaciers there and in Greenland have been galloping into the sea. . . . And the speeding glaciers, at least, are surely driving up sea level and pushing shorelines inland. . . . Some of the glaciers draining the great ice sheets of the Antarctic and Greenland have speeded up dramatically, driving up sea level and catching scientists unawares. They don't fully understand what is happening. And if they don't understand what a little warming is doing to the ice sheets today, they reason, what can they say about the ice's fate and rising seas in the greenhouse world of the next century or two?

The freshening of the North Atlantic Ocean, the slowdown of the Atlantic thermohaline circulation, the melting of alpine glaciers, and the melting of the Greenland and West Antarctic ice sheets constitute the "smoking gun" of global warming. "We are seeing dangerous, human-induced climate change," says Michael Schlesinger, author of *Avoiding Dangerous Climate Change*, "The melting of the Greenland ice sheet would raise sea level by 18 feet. Melting of the Antarctic ice sheet would raise sea level an additional 22 feet. Most coastal cities would be inundated. These observed changes in climate and ongoing research have shown that human-induced warming is proceeding more quickly than anticipated. Schlesinger, professor of atmospheric sciences at the University of Illinois at Urbana-Champaign, continues, "Not only are the Greenland and Antarctica ice sheets melting much faster than models predicted, measurements show a significant freshening (influx of fresh water) of the North Atlantic Ocean and a 30 percent reduction of North Atlantic circulation within the past 50 years."

Naturally, high-seas fisheries will be less affected by global warming than those on the coasts and inland. Nevertheless, oceanographers are currently unable to rule out the possibility that global warming may exacerbate hurricanes, typhoons, cyclones, El Niños, and other climate fluctuations. One of the suggested causes of the proliferation of Western Atlantic hurricanes during the 2005 season was the warming of the sea surface, which MIT meteorologist Kerry Emmanuel says "may lead to an upward trend in tropical cyclone destructive potential." Emmanuel's article "Increasing Destructiveness of Tropical Cyclones over the Past 30 Years" was published in *Nature* on August 4, 2005, three weeks before Hurricane Katrina hit New Orleans. Although their study did not include the catastrophic Katrina, Carlos Hoyos and colleagues (of the School of Earth and Atmospheric Sciences, Georgia Institute of Technology), published an important paper (in a 2006 issue of *Science* in which they note "that the trend of category 4 and 5 hurricanes for the period 1970–2004 is directly linked to the trend in sea-surface temperature."

To the benefit of fishermen and the detriment of fish, warmer temperatures are likely to enhance fishing productivity. Overall biological activity is greater at higher temperatures; as more food is available, fish grow faster, and they reproduce at a younger age. The expected increase in fisheries from warmer temperatures may be partly offset, however, by a decline in upwellings, the upward flow of deep ocean water to the surface. These upward flows bring nutrients to the upper layers of the ocean, increasing the growth of the plankton that forms the base of the marine food chain. The swordfish, whose regular habits include movements toward cold waters for feeding and to warm waters for spawning and overwintering, will definitely be affected if the cold waters are warmed, even by a small amount.

Throughout tropical oceans, oxygen-poor regions are expanding as the oceans warm up. This is bad news for predatory fishes because the microooganisms at the base of the food pyramid cannot survive in low-oxygen (hypoxic) zones. In a 2008 report, Stramma et al. predicted declines in oceanic dissolved oxygen produced by global warming, which may have dramatic consequences for the large predatory fish species. In a study published in 2011, NOAA fisheries biologist Eric Prince and colleagues found that expanding ocean "dead zones"—areas where oxygen levels are so low that creatures cannot survive over the long term—are shrinking the habitat for high-value fish such as marlins in the tropical northeast Atlantic Ocean. Dead zones, or hypoxic areas in the world's oceans, have been on the increase since oceanographers began observing them in the 1970s. These usually occur near inhabited coastlines, where aquatic life is most concentrated, but rarely in the vast middle portions of the oceans, which naturally have little life. As dead zones expand, marlins, swordfish, other billfish, and tunas—high-energy fish that need large amounts of dissolved oxygen—move into surface waters where they are more vulnerable to fishing. Prince et al.'s paper focuses on the tropical northeast Atlantic Ocean off Africa, but the expansion of low-oxygen zones is occurring in all tropical ocean basins and throughout the subarctic Pacific, making the compression of habitat a global issue. This paper follows earlier research by Prince and colleagues published in 2010 in *Fisheries Oceanography* based on tagging of marlins and sailfish in the waters off Florida and the Caribbean, which also showed that these billfish prefer oxygen-rich waters close to the surface and that they move away from waters low in dissolved oxygen.

Because we have so fouled our terrestrial nest, we want to believe that the ocean is still some sort of deep, safe haven for its inhabitants. Isn't it? In a word: no. We have spilled millions of gallons of oil at various locations; we have dumped sewage and toxic wastes into the deep; we have littered the ocean from surface to floor with every kind of non-biodegradable refuse; and of course, we have "harvested" some of the ocean's inhabitants in such prodigious numbers that we have—some say irrevocably—altered the balance of life in the sea. Coral reefs are threatened around the world, but so far, we have not conclusively identified any recent marine extinctions because it is impossible to say that there is not a single barndoor skate or white abalone left somewhere in the ocean. There may indeed by many more extinct marine species than we know about because underwater fossils are much scarcer than terrestrial ones—or at least harder to spot.

The past history of the ocean is punctuated by extinctions on every scale. The list of extinct invertebrates, marine mammals, reptiles, and fishes is so extensive that it is possible only to mention the most noteworthy ones: the trilobites, the ammonites, the belemnites, jawless fishes, proto-whales, early sharks (including the megapredator *Megalodon*), and every prehistoric marine reptile, including the ichthyosaurs, mosasaurs, and plesiosaurs. (Of the early marine reptiles, only the sea turtles have survived, and their days are surely numbered.) The recent disappearance of some animal species, however, presents no mystery at all.[1]

We know precisely why there are no more dodos, giant sea cows, great auks, Caribbean monk seals, passenger pigeons, Tasmanian wolves, Eskimo curlews, and Carolina parakeets. We destroyed their habitat so they had no place to live, or we killed them, every last one. But except for those extirpations where we have but to look in a mirror to identify the cause, there is no inclusive theory as to why extinction happens. The plesiosaurs and ichthyosaurs are long gone, as are the trilobites, ammonites, belemnites, and millions of other lesser-known marine life forms. What happened to them? We don't know. We do know, however, that throughout Earth's long history, certain events, characterized as "mass extinctions," have occurred, and on some available evidence, we are able to speculate as to the causes of these events, which saw the disappearance of so many species.

Some 250 million years ago—give or take a couple of million either way—nearly all life on Earth was wiped out. Sometimes known as the Mother of all Extinctions, the Permian extinction took out 95 percent of all living things, making it the greatest biological catastrophe of all time. Until recently, paleontologists believed that the Permian event lasted for five to eight million years and was associated primarily with

1. As we've seen, the white marlin (now *Kajikia albidus*) was poised on the brink of endangerment because of overfishing. Though a DNA analysis of many of these fishes showed they were not white marlins but, rather, roundscale spearfishes and the categorization was then modified, there are an awful lot of these fishes being caught, regardless.

massive global changes, such as the tectonic activity that brought about the new continent of Pangaea and produced earth-shattering changes in climate and sea level. As the plates moved and ground against each other, perhaps magma exploding through the earth's crust released enough gases to blot out the sun for years, depriving plants of the ability to photosynthesize. If so little oxygen was pumped into the atmosphere, CO_2 and other greenhouse gases would accumulate and raise global temperatures enough to affect every living thing. In his 1993 book about the Permian extinction, Smithsonian paleobiologist Douglas Erwin writes:

> Undoubtedly the most important geologic event in the Permian was the formation of the great supercontinent Pangea This land mass included virtually all of the large continental fragments, although some microplates never collided with Pangea. The southern continents (what are now South America, Africa, India, Antarctica, Australia, and parts of the Middle East and southeast Asia) were united into a large continent known as Gondwana during the Paleozoic. North of Gondwana lay Laurasia, which included North America, Europe, the Russian and Siberian platforms, and Kazakhstan.

There are those who suggest that it happened rapidly, in less than half a million years. In a study published in 2000, several scientists from the Nanjing Institute of Geology and Paleontology (Jin and colleagues, including Douglas Erwin) suggest that the event took place almost overnight (geologically speaking), "coincident with the eruption of the Siberian flood basalts." In 500,000 years, volcanoes that spread over Siberia from the Ural Mountains to Lake Baikal poured out lava that covered the land in a blanket two miles thick. The lava would have killed everything it covered, but much more damaging would have been the debris and noxious gases (such as CO_2) generated by the ongoing eruptions. But by some researchers' calculations, the Permian extinction happened even faster, meaning that not even volcanic eruptions could explain the widespread annihilation of life, and they continue to search for evidence of an extraterrestrial impact.

There is strong evidence that the end of the Permian was characterized by massive volcanic eruptions, particularly in the area known as the Siberian Traps, causing major disruptions in the atmosphere and covering 2.5 million square miles of what is now eastern Russia in mile-thick lava. In 1997, geologists Antoni Hallam and Paul Wignall published *Mass Extinctions and Their Aftermath*, in which they showed how the eruption of the Siberian Traps would have caused acid rain, high emissions of carbon dioxide and sulfur dioxide, global warming, global darkness, and eventually, mass extinction. Benton concludes: "It took 20 or 30 million years for coral reefs to reestablish themselves, and for the forests to re-grow. In some settings, it took 50 million years for full ecosystem complexity to recover. Geologists and paleontologists are only beginning to get to grips with this most profound of crises." Michael Benton himself wrote a book about the Permian extinction, which he called *When Life Nearly Died: The Greatest Mass Extinction of All Time*. He argues that the event was not initiated by an extraterrestrial agent at all but, rather, by massive volcanism, which released methane gases into the sea and poisonous gases into the atmosphere, snuffing out life forms in both biomes.

On August 12, 1986, a cloudy mixture of carbon dioxide (CO_2) and water droplets rose violently from Lake Nyos in Cameroon, western Africa. Normally, CO_2 remains in solution—as with the bubbles in an unopened bottle of soda—but something forced the gas to the surface (the lake sits atop an old volcano) and it rose and burst from the surface releasing a deadly cloud of heavier-than-air, poisonous gas. Thousands of cattle, innumerable other animals, and 1,800 people were asphyxiated. These deaths hardly count as a mass extinction, but the event was one of the factors that prompted Gregory Ryskin of Northwestern University to suggest that a massive eruption of methane gas (CH_4) from the ocean might have exploded and incinerated the land dwellers of the Permian. In his 2003 article in *Geology*, he writes:

> The consequences of a methane-driven oceanic eruption for marine and terrestrial life are likely to be catastrophic. Figuratively speaking, the erupting region "boils over," ejecting a large amount of methane and other gases (e.g., CO_2, H_2S [hydrogen sulfide]) into the atmosphere, and flooding large areas of land. Whereas pure methane is lighter than air, methane loaded with water droplets is much heavier, and thus spreads over the land, mixing with air in the process (and losing water as rain). The air-methane mixture is

explosive at methane concentrations between 5% and 15%; as such mixtures form in different locations near the ground and are ignited by lightning, explosions and conflagrations destroy most of the terrestrial life, and also produce great amounts of smoke and carbon dioxide.

Such a methane explosion might have brought on the Permian extinction by killing the land dwellers and also killing the land plants. "The paleogeography of the Permian," writes Ryskin, "may have led to the development of a large number of stagnant anoxic regions, and thus to accumulation of very large amounts of dissolved methane. The unusual severity of the Permian-Triassic extinction may have been the result of chance as several different oceanic locations erupted in succession," or what extinction theorist David Raup categorizes as "bad luck." Sooner or later, every living species is destined for extinction.

The most notorious and numerous victims of the luck of the Permian event (or events) were the trilobites, which perdured for 150 million years longer than the dinosaurs, but became just as extinct. They were underwater arthropods ("joint-legs") whose origins can be traced as far back as 540 million years ago, to the Middle Cambrian Period. From the plentiful fossil evidence, we know that there were more (perhaps many more) than 15,000 distinct species, ranging in length from a millimeter to a foot. They had well-developed eyes—believed to be the first such organs—and while most species seemed to crawl along the bottom, there were some that were evidently free-swimming. These varied creatures occupied the oceans of the world for approximately 290 million years—an evolutionary success story by any definition—but when the waters had cleared from the effects of the Permian extinction, the trilobites were gone.

Gregory Retallack, Roger Smith, and Peter Ward think that the release of oceanic gases might have contributed to the Permian extinction but not by igniting and burning up the land and all the animal life. Methane, they suggest, instead, would have reduced the relative amount of oxygen in the air, which would have triggered pulmonary and cerebral edema, the same thing that affects mountaineers at high altitudes. Land animals would have suffocated, and marine life would have been similarly affected by oxygen-poor water. Whether the cause was extraterrestrial impact, massive volcanic lava flows, methane eruptions, or some combination of these, the earth and the ocean heated up under a blanket of greenhouse gases, and an almost incomprehensible number of living things were killed off, never to propagate again.

The Permian mass extinction occurred about 250 million years ago and was the greatest mass extinction ever recorded in earth history—significantly larger than the Ordovician and Devonian crises and much more devastating than the better-known K-T extinction that felled the dinosaurs. In the 2006 revision of his 1993 book (now called, elegantly, *Extinction*), Douglas Erwin admits again that we still don't know the nature of the global catastrophe that almost wiped out all life in earth. "Among the myriad suggestions," he writes, "four stand out as possible causes: the impact of a meteorite or comet; climatic destruction from massive volcanism in Siberia; the oceans losing their oxygen and snuffing out the animals that required it; and a combination of several interacting and mutually reinforcing events." He continues:

> Almost everything we know is consistent with an extinction caused by the collision of an extraterrestrial object at the very end of the Permian, a very rapid extinction. [There was a] dramatic shift in the flow of carbon through the oceans and atmosphere, and extinction on land and ocean . . . [but] there is no sign of extraterrestrial elements such as iridium. Some geologists have offered tantalizing suggestions of impact, but so far these have failed to convince most scientists. . . . Geologists have uncovered considerable evidence for anoxic [low or no oxygen] waters in both the deep sea and shallow water near the Permo-Triassic boundary, and this has led to the third hypothesis: the spread of low-oxygen waters. . . . If anoxia was the major cause of extinction [in the oceans], it must have been linked to some other process responsible for extinctions on land.

The primary marine and terrestrial victims of the Permian extinction included the fusulinid foraminifera, trilobites, rugose and tabulate corals, blastoids, acanthodians, placoderms, and pelycosaurs. Other groups that were substantially reduced included the bryozoans, brachiopods, ammonoids, sharks, bony fish, crinoids, eurypterids, ostracods, and echinoderms. In *The Great Paleozoic Crisis*, his detailed study of the Permian extinction, Douglas Erwin writes (after comparing this mystery to Agatha Christie's *Murder on the Orient Express*, where *all* the suspects had a hand in the murder):

> I believe that the extinction cannot be traced to a single cause, but rather a multitude of events occurring together, in particular the increased climatic and ecologic instability associated with the regression and a combination of greenhouse warming and possible oceanic anoxia from increased atmospheric CO_2. . . . The major phase of the marine extinction appears to have begun earlier on the mainland of Pangea with the onset of the regression and the corresponding loss in habitat diversity and reduction of shelf area. The major extinctions were triggered by the dramatic increase in CO_2 and methane from oxidation of organic material, gas hydrates, and possibly other sources.

By 2006, however, Erwin had more or less rejected the Agatha Christie solution, only because it was "the most difficult to test." Some of the suggested causes can be examined independently, but it is nigh onto impossible to program all of the factors present 250 million years ago and come up with anything but chaotic results.

Steven M. Stanley, an evolutionary biologist at Johns Hopkins University, believes that climate change is the primary agent for marine extinctions. In a chapter in *his* book *Extinction*, he elaborates thus:

> One of the great, long-standing puzzles of paleontology has been what caused the great biotic crises that we call mass extinctions. . . . I present evidence bearing on the relative roles of two environmental changes frequently cited as potential agents of marine mass extinctions: regression of shallow seas and worldwide temperature change. I conclude that reduction of living space due to regression of shallow seas has been of trivial importance, whereas temperature change has served as a prominent agent of mass extinction.

For most of the mass extinctions, Stanley identifies climatic *cooling* as the cause. Moreover, by definition, extinctions by climate change—as contrasted, say, to anthropogenic extinctions or collisions with extraterrestrial objects—take a very long time indeed. Even though it seems recent, the idea of human-induced global warming has been around for some time. In 1895, Swedish chemist Svante Arrhenius (1859–1927) presented a paper to the Stockholm Physical Society titled *On the Influence of Carbonic Acid* [H_2CO_3, carbon dioxide dissolved in water] *in the Air on the Temperature of the Ground*, in which he argued that combustion of fossil fuel would lead to global warming. He knew that carbon dioxide absorbs infrared radiation resulting in a warming effect on the atmosphere and calculated that the surface of the earth was about 30°C warmer than it would be if there were no atmosphere. In 1956, according to Spencer Weart's *The Discovery of Global Warming*, a physicist named Gilbert Plass identified the greenhouse effect and announced that human activity would raise the average global temperature "at the rate of 1.1 degree C per century." With the possible exception of polar bears, no living species appears to be immediately threatened by the warming of the world's oceans, but there is no question that warmer oceans can pose a threat to continuation of a species; the Permian extinction reminds us that warmer seas have been implicated in the disappearance of more than a few species.

Douglas Erwin does not think that early mass extinctions should be used to predict modern extinctions because "any comparison of fossil extinction rates to current estimates is inherently flawed because the data are so different." I am not suggesting that an increase of a couple of degrees in sea-surface temperature will produce anything like a mass extinction, but it stands to reason that creatures already sensitive to temperature fluctuations—swordfish, for example—might be adversely affected if the ocean warms up. It would be the first time in the planet's history that a single species (us) might be responsible for the climate change that eradicated another.

In her 2000 study of the rise and fall of the Cape Breton (Nova Scotia) swordfishery (the location where Michael Lerner took the AMNH expedition in 1936), Gretchen Fitzgerald asserts, "It is common knowledge amongst swordfishermen that the distribution of swordfish in Canadian waters is highly dependent on temperature. Harpooners look for water that is approximately 60°F or 15°C, where fish are known to come to the surface. The harpooners also need a number of clear, calm days or 'greasers' in the summertime so that fish can be spotted when they fin. . . . Because swordfish distribution is highly dependent upon temperature, it seems probable that changes in the path of the Gulf Stream and in ocean temperatures could produce changes in the distribution and number of swordfish observed off our coasts." At first, warmer waters will encourage breeding, and the encouragement of breeding could produce more swordfish. If the swordfish proliferate, it might encourage more swordfishing, which will then begin the cycle of overfishing followed by a population crash all over again.

Xiphias the swordfighter has managed to survive thousands of years of netting, harpooning, rod-and-reeling, long-lining, and drift-netting. It has survived charcoal grilling, baking, microwaving, and sashimi making. At least one swordfish survived an encounter with Mark Ferrari—although Ferrari nearly didn't survive. They have not, however, survived encounters with submersibles, dories, ships, whales, and bales of rubber. Although the swordfish received a reprieve when it appeared that overfishing was directing the species toward commercial extinction, it remains to be seen if the great fish can survive into the next century, as the earth's support systems—the ocean, the atmosphere, and the weather—are modified. Think of cities, dams, highways, bridges, strip mines, garbage dumps, irrigation ditches, clear-cut forests, smog, acid rain, pollutant aerosols, napalm, and nuclear bomb tests. In an essay written in 1967—*1967!*—Lynn White stated,

> When the first cannons were fired, in the early 14th century, they affected ecology by sending workers scrambling to the forests and mountains for more potash, sulfur, iron ore, and charcoal, with some resulting erosion and deforestation. Hydrogen bombs are of a different order: a war fought with them might alter the genetics of all life on this planet. By 1285 London had a smog problem arising from the burning of soft coal, but our present combustion of fossil fuels threatens to change the composition of the globe's atmosphere as a whole, with consequences that we are only beginning to guess. With the population explosion, the carcinoma of planless urbanism, the now geological deposit of sewage and garbage, surely no creature other than man has ever managed to foul its nest in such a short time.

We modify the earth, the oceans, the atmosphere, and the weather at our peril—and the peril of all life on earth.

Bibliography

Adler, Jerry. "Health: Is Salmon as Good for You as You Think?" *Newsweek*, October 28, 2002, 54–55.

Aelian. *Aelian on the Characteristics of Animals*. Ed. and trans. A. F. Scholfield. Loeb Classical Library. Cambridge, MA: Harvard University Press, 1959.

Akyol, O., and T. Ceyhan. "The Turkish Swordfish Fishery." *Collective Volume of Scientific Papers ICCAT* 66, no. 4 (2011): 1471–79.

Argüelles, J., P. G. Rodhouse, P. Villegas, and G. Castillo. "Age, Growth and Population Structure of the Jumbo Flying Squid *Dosidicus gigas* in Peruvian Waters." *Fisheries Research* 54 (2001): 51–61.

Aristotle. *Historia animalium*. Ed. Allan Gotthelf. Trans. D. M. Balme and A. L. Peck. Loeb Classical Library. Cambridge, MA: Harvard University Press, 1965–91.

Arnold, S. M., T. V. Lynn, L. A. Verbrugge, and J. P. Middaugh. "Human Biomonitoring to Optimize Fish Consumption Advice: Reducing Uncertainty When Evaluating Benefits and Risks." *American Journal of Public Health* 95, no. 3 (2005): 393–97.

Arocha, F. "Implicaciones de Ordenación Pesquera en el Pez Espada *Xiphias gladius*, del Atlantico Noroccidental." *Boletín del Instituto Oceanográfico de Venezuela*, 57, nos. 1–2 (1998): 81–89.

Arocha, F., and D. E. Lee. "Maturity at Size, Reproductive Seasonality, Spawning Frequency, Fecundity and Sex Ratio in Swordfish from the Northwest Atlantic." *Collective Volume of Scientific Papers ICCAT* 45, no. 2 (1996): 350–57.

———. "Preliminary Observations on Sex Ratio and Maturity Stages of the Swordfish *Xiphias gladius*, in the Northwest Atlantic." *Collective Volume of Scientific Papers ICCAT* 40, no. 1 (1993): 126–31.

———. "The Spawning of Swordfish from the Northwest Atlantic." *Collective Volume of Scientific Papers ICCAT* 44, no. 3 (1995): 179–86.

Arrhenius, S. "On the Influence of Carbonic Acid in the Air Upon the Temperature of the Ground." *Philosophical Magazine* 41 (1896): 237–76.

Bandini, R. "Swordfishing." *California Fish and Game* 19 (1933): 241–48.

———. *Veiled Horizons: Stories of Big Game Fish of the Sea*. Lyon, MS: Derrydale Press, 1939.

Barnard, J. "Salmon Woes Linked to Weather." *Seattle Times*, March 4, 2008.

Barnes, D. "The World's Great Billfish." *Field and Stream* 92, no. 11 (1988): 77–85.

Battersby, S. "Deep Trouble." *New Scientist* 190 (2006): 42–46.

Beale, T. *A Few Observations on the Natural History of the Sperm Whale.* London: Wilson, 1835.

Beardsley, G. L. "Report of the Swordfish Workshop Held at the Miami Laboratory, Southeast Fisheries Center, Miami, Florida, June 7–9, 1977." *Collective Volume of Scientific Papers ICCAT* 7, no.1 (1978): 149–58.

Beardsley, G. L., R. J. Conser, A. M. Lopez, M. Brassfield, and D. McLellan. "Length and Weight Data for Western Atlantic Swordfish, *Xiphias gladius*." *Collective Volume of Scientific Papers ICCAT* 8, no. 2 (1979): 490–95.

Bearzi, G., E. Politi, S. Agazzi, and A. Azzellino. "Prey Depletion Caused by Overfishing and the Decline of Marine Megafauna in the Eastern Ionian Sea Coastal Waters (Central Mediterranean.)" *Biological Conservation*127 (2006): 373–582.

Beerkircher, L., F. Arocha, A. Barse, E. Prince, V. Restrepo, J. Serafy, and M. Shivji. "Effects of Species Misidentification on Population Assessment of Overfished White Marlin *Tetrapus albidus* and Roundscale Spearfish *T. georgii*." *Endangered Species Research* 9 (2009): 81–90.

Bello, G. "Role of Cephalopods in the Diet of the Swordfish, *Xiphias gladius*, from the Eastern Mediterranean Sea." *Bulletin of Marine Science* 49, nos. 1–2 (1991): 312–24.

Bennett, F. D. *Narrative of a Whaling Voyage around the Globe from the Year 1833 to 1836 . . . with an Account of the Southern Whales, the Sperm Whale Fishery, and the Natural History of the Climates Visited.* London: Richard Bentley, 1840.

Benton, M. J. *When Life Nearly Died: The Greatest Mass Extinction of All Time.* New York: Thames and Hudson, 2003.

Benton, M. J., and R. J. Twichett. "How to Kill (Almost) All Life: The End-Permian Extinction Event." *Trends in Ecology and Evolution* 18, no. 7 (2003): 358–65.

Bigelow, H. B., and W. C. Schroeder. "Fishes of the Gulf of Maine." *US Fish and Wildlife Service Fisheries Bulletin* 74 (1953): 1–577.

Block, B. A. "Physiology and Ecology of Brain and Eye Heaters in Billfishes." In *Planning the Future of Billfishes: Research and Management in the 90s and Beyond.* Pt. 2: *Contributed Papers*, edited by R. H. Stroud, 123–36. International Billfish Symposium, Kailua-Kona, Hawaii, August 1–5, 1988. Savannah, GA: National Coalition for Marine Conservation, 1990.

Block, B. A., D. Booth, and F. G. Carey. "Direct Measurement of Swimming

Speeds and Depth of Blue Marlin." *Journal of Experimental Biology* 166 (1992): 267–84.
Block, B. A., H. Dewar, C. Farwell, and E. D. Prince. "A New Satellite Technology for Tracking the Movements of Atlantic Bluefin Tuna." *Proceedings of the National Academy of Sciences* 95, no. 16 (1998): 9384–89.
Brill, R. W. "Selective Advantages Conferred by the High Performance Physiology of Tunas, Billfishes, and Dolphin Fish." *Comparative Biochemical Physiology* 113A, no. 1 (1996): 3–15.
Broad, W. J. "The Storied Narwhal Begins to Reveal the Secrets of Its Tusk." *New York Times*, December 13, 2005, F1–F4.
Bromhead, D., J. Pepperell, B. Wise, and J. Findlay. *Striped Marlin: Biology and Fisheries*. Canberra: Bureau of Rural Sciences, 2003.
Brownlee, J. "Return of the Swordfish." *Saltwater Sportsman* 63, no. 8 (2002): 26–33.
Bruemmer, F. "The Sea Unicorn." *Audubon* 71, no. 6 (1969): 58–63.
Buel, J. W. *Sea and Land: An Illustrated History of the Wonderful and Curious Things of Nature Existing before and since the Deluge: A Natural History of the Sea, Land Creatures, the Cannibals, and Wild Races of the World.* Philadelphia: Historical Publishing Co., 1887.
Bullen, F. T. *Denizens of the Deep*. New York: Fleming H. Revell, 1904.
Bundy, A. "Fishing on Ecosystems: The Interplay of Fishing and Predation in Newfoundland-Labrador." *Canadian Journal of Fisheries and Aquatic Sciences* 58, no. 6 (2001): 1153–67.
Byatt, A., A. Fothergill, and M. Holmes. *The Blue Planet*. London: BBC Worldwide Limited, 2001.
Campbell, R. A., J. G. Pepperell, and T. L. O. Davis. "Use of Charter Boat Data to Infer the Annual Availability of Black Marlin, *Makaira indica*, to the Recreational Fishery off Cairns, Australia." *Marine and Freshwater Research* 54, no. 4 (2003): 447–57.
Canada. Fisheries Branch. *Annual Report of the Department of Marine and Fisheries*. Ottawa: Department of Marine and Fisheries, 1903.
Canese, S., F. Garibaldi, L. Orsi Relini, and S. Greco. "Swordfish Tagging with Pop-Up Satellite Tags in the Mediterranean Sea." *Collective Volume of Scientific Papers ICCAT* 62, no. 4 (2008): 1052–57.
Capozzi, E. "The First Billfishermen." *Marlin*, January 2007, 56–59.
Card, I. "Out of the Blue." As told to Colin Kearns. *SaltWater Sportsman* 67, no. 11 (2006): 84–86.
Carey, F. G. "A Brain Heater in the Swordfish." *Science* 216 (1982): 1327–29.
———. "Fishes with Warm Bodies." *Scientific American* 228, no. 2 (1973): 36–44.

——. "Further Acoustic Telemetry Observations of Swordfish." In *Planning the Future of Billfishes: Research and Management in the 90s and Beyond*. Pt. 2: *Contributed Papers*, edited by R. H. Stroud, 103–22. International Billfish Symposium, Kailua-Kona, Hawaii, August 1–5, 1988. Savannah, GA: National Coalition for Marine Conservation, 1990.

——. "Through the Thermocline and Back Again." *Oceanus* 35, no. 3 (1992): 79–85.

Carey, F. G., and B. H. Robison. "Daily Patterns in the Activities of Swordfish *Xiphias gladius*, Observed by Acoustic Telemetry." *Fishery Bulletin* 79, no. 2 (1981): 277–92.

Carey, F. G., and J. M. Teal. "Heat Conservation in Tuna Fish Muscle." *Proceedings of the National Academy of Sciences* 56, no. 5 (1966): 1464–69.

Carson, R. L. "Food from the Sea: Fish and Shellfish of New England." *US Department of the Interior Conservation Bulletin* 33 (1943): 1–74.

Christensen, V., S. Guénette, J. J. Heymans, C. J. Walters, R. Watson, D. Zeller, and D. Pauly. "Estimating Fish Abundance of the North Atlantic, 1950 to 1999." In *Fisheries Impacts on North Atlantic Ecosystems: Models and Analyses*, edited by S. Guénette, V. Christensen, and D. Pauly, 1–25. Fisheries Centre Research Reports 9, no. 4. Vancouver: Fisheries Centre, University of British Columbia, 2002.

Church, R. L. "Broadbill Swordfish in Deep Water." *Sea Frontiers* 14, no. 4 (1968): 246–49.

——. "*Deepstar* Explores the Ocean Floor." *National Geographic* 139, no. 1 (1971): 110–29.

Clarke, G. L., and R. H. Backus. "Interrelations between the Vertical Migrations of Deep-Scattering Layers, Bioluminescence, and Changes in Daylight in the Sea." *Bulletin of the Institute of Oceanography of Monaco* 64, no. 1318 (1964): 1–36.

——. "Measurement of Light Penetration in Relation to Vertical Migration and Records of Luminescence in Deep-Sea Animals." *Deep-Sea Research* 4 (1956): 1–14.

Clarke, M. R. *A Handbook for the Identification of Cephalopod Beaks*. Oxford: Clarendon Press, 1986.

Clarkson, T. W. "Human Toxicology of Mercury." *Journal of Trace Elements in Experimental Medicine* 11, nos. 2–3 (1998): 303–17.

——. "The Three Modern Faces of Mercury." *Environmental Health Perspectives* 110, no. S1 (2002): 11–23.

Clarkson, T. W., and J. J. Strain. "Nutritional Factors May Modify the Toxic Action of Methyl Mercury in Fish-Eating Populations." *Journal of Nutrition* 133 (2003): 1539–43.

Clayton, M. "New Questions about Safety of Tuna Imports." *Christian Science Monitor*, July 12, 2006.

Clover, C. *The End of the Line: How Overfishing Is Changing the World and What We Eat*. New York: New Press, 2006.

Coghlan, A. "Extreme Mercury Levels Revealed in Whalemeat." *New Scientist* 173 (2002): 11.

———. "Shops in Japan Are Selling Mercury-Ridden Dolphin Flesh as Whalemeat." *New Scientist* 178 (2003): 7.

Cole, K. S., and D. L. Gilbert. "Jet Propulsion of Squid." *Biological Bulletin* 138, no. 3 (1970): 245–46.

Collette, B. B., J. R. McDowell, and J. E. Graves. "Phylogeny of Recent Billfishes." *Bulletin of Marine Science* 79, no. 3 (2006): 455–68.

Cone, M. "Warning on Tuna Cans Is Rejected." *Los Angeles Times*, May 13, 2006.

Conrad, G. M. "The Nasal Bone and Sword of the Swordfish (*Xiphias gladius*)." *American Museum Novitates* 968 (1937): 1–3.

Conrad, G. M., and F. R. LaMonte. "Observations on the Body Form of the Blue Marlin (*Makaira nigricans ampla* Poey)." *Bulletin of the American Museum of Natural History* 74, no. 4 (1937): 208–20.

Cope, E. D. "[On an Extinct Genus of Saurodont Fishes.]" *Proceedings of the Academy of Natural Sciences of Philadelphia* 24: (1873) 280–81.

———. "On the Two New Species of Saurodontidae." *Proceedings of the Academy of Natural Sciences of Philadelphia* 25 (1873): 337–39.

Cornell, C. "Harpooning Swordfish on Georges Bank." *National Fisherman* 62, no. 9 (1982): 12–14.

———. "Harpooning Trip to Georges Ends in Success despite Fog and Wind." *National Fisherman* 62, no. 10 (1982): 26–28, 110.

Cramer, J. "Effect of Regulations Limiting Landings of Swordfish by Weight on Commercial Pelagic Longline Fishing Patterns." In *Fisheries Bycatch: Consequences and Management*, 63–64. Alaska Sea Grant College Program Report, no. 97-02. Fairbanks: Alaska Sea Grant College Program, University of Alaska, 1996.

———. "Pelagic Longline Bycatch." *Collective Volume of Scientific Papers ICCAT* 55, no. 4 (2003): 1576–86.

Croker, R. S. "Further Notes on the Giant Squid, *Dosidicus gigas*." *California Fish and Game* 23, no. 3 (1937): 246–47.

Curry, R., B. Dickson, and I. Yashaeyev. "A Change in the Freshwater Balance of the Atlantic Ocean over the Past Four Decades." *Nature* 426 (2003): 826–29.

Dagorn, L., F. Menczer, P. Bach, and R. J. Olson. "Co-evolution of Movement Behaviours by Tropical Pelagic Predatory Fishes in Response to Prey

Environment: A Simulation Model." *Ecological Modeling* 134 (2000): 325–41.

Dalzell, P., and C. H. Boggs. "Pelagic Fisheries Catching Blue and Striped Marlins in the US Western Pacific Islands." *Marine and Freshwater Research* 54, no. 4 (2003): 419–24.

Davenport, D., J. R. Johnson, and J. Timbrook. "The Chumash and the Swordfish." *Antiquity* 67 (1993): 257–72.

Davidson, A. *North Atlantic Seafood*. New York: Harper & Row, 1980.

Davidson, P. W., G. J. Meyers, C. Cox, C. Axtell, C. Shamlaye, J. Sloan-Reeves, E. Cernichiari, et al. "Effects of Prenatal and Postnatal Methylmercury Exposure from Fish Consumption on Neurodevelopment: Outcomes at 66 Months of Age in the Seychelles Development Study." *Journal of the American Association of Medicine* 280, no. 8 (1998): 701–7.

Davis, J. E. "Not a Fish Story." *San Francisco Chronicle*, August 3, 2005, B9.

DeMartini, E. E., J. H. Uchiyama, and H. A. Williams. "Sexual Maturity, Sex-Ratio, and Size Composition of Swordfish, *Xiphias gladius*, Caught by the Hawaii-Based Pelagic Longline Fishery." *Fishery Bulletin* 98, no. 3 (2000): 489–506.

DeMetrio, G., H. Ditrich, and G. Palmieri. "Heat-Producing Organ of the Swordfish (*Xiphias gladius*): A Modified Eye Muscle." *Journal of Morphology* 234, no. 1 (1997): 89–96.

de Sylva, D. P. "Juvenile Blue Marlin, *Makaira ampla* (Poey) from Miami, Florida, and West End, Bahamas." *Bulletin of the American Museum of Natural History* 114, no. 5 (1958): 412–15.

——. "Postlarva of the White Marlin, *Tetyrapturus albidus*, from the Florida Current off the Carolinas." *Bulletin of Marine Sciences of the Gulf and Caribbean* 13, no. 1 (1963): 123–32.

Devlin, J. C. "Swordfish Duels Two-Man Research Submarine." *New York Times*, January 14, 1968, 15.

Dewhurst, W. H. *The Natural History of the Order Cetacea and the Oceanic Inhabitants of the Arctic Regions*. London, 1835.

Diamond, J. M. "The Present, Past, and Future of Human-Caused Extinctions." In *Evolution and Extinction*, edited by W. G. Chaloner and A. Hallam, 229–37. Cambridge: Cambridge University Press, 1989.

Dibenedetto, D. "When Marlin Strike Back." *Saltwater Sportsman* 66, no. 9 (2005): 14.

Di Natale, A., A. Celona, and A. Mangano. "A Series of Catch Records by the Harpoon Fishery in the Strait of Messina from 1976 to 2003." *Collective Volume of Scientific Papers ICCAT* 58, no. 4 (2005): 1348–59.

Di Natale, A., A. Mangano, A. Asaro, M. Bascone, A. Celona, E. Navarra, and M. Valastro. "Swordfish (*Xiphias gladius* L.) Catch Composition in the Tyrrhenian Sea and in the Straits of Sicily in 2002 and 2003." *Collective Volume of Scientific Papers ICCAT* 58, no. 4 (2005): 1511–36.

Di Natale, A., A. Mangano, A. Celona, and M. Valastro. "Size Frequency Composition of the Mediterranean Spearfish (*Tetrapturus belone*, Rafinesque) Catches in the Tyrrhenian Sea and the Strait of Messina in 2003." *Collective Volume of Scientific Papers ICCAT* 58, no. 2 (2005): 589–95.

Di Natale, A., A. Mangano, A. Maurizi, I. Montaldo, E. Navarra, S. Pinca, G. Schimmenti, G. Torchia, and M. Valastro. "A Review of Catches by the Italian Fleet: Species Composition, Observers Data and Distribution along the Net." *Collective Volume of Scientific Papers ICCAT* 44, no. 1 (1995): 226–35.

Domier, M. L., H. Dewar, and N. Nasby-Lucas. "Mortality Rate of Striped Marlin (*Tetrapterus audax*) Caught with Recreational Tackle." *Marine and Freshwater Research* 54, no. 4 (2003): 435–45.

dos Santos, M. N., and A. Garcia. "The Influence of the Moon Phase on the CPUES for the Portuguese Swordfish (*Xiphias gladius* L., 1758) Fishery." *Collective Volume of Scientific Papers ICCAT* 58, no. 4 (2005): 1466–69.

Doumas, C. *The Wall Paintings of Thera*. Athens: Thera Foundation, 1992.

Doyle, A. "Norway Advises Pregnant Women against Whale Meat." Reuters, May 15, 2003. http://reuters.com/newsArticle.jhtml?type=topNews&storyID +27226774.

Duncan, D. D. "Fighting Giants of the Humboldt." *National Geographic* 79, no. 3 (1941): 373–400.

Dunn, B., and P. Goadby. "When Swordfish Attacked Boats." *Classic Angling* 3 (1999): 18–21.

Ehrhardt, N. M. "Age and Growth of Swordfish *Xiphias gladius* in the Northwestern Atlantic." *Bulletin of Marine Science* 50, no. 2 (1992): 292–301.

Ehrhardt, N. M., R. J. Robbins, and F. Arocha. "Age Validation and Growth of the Swordfish, *Xiphias gladius*, in the Northwest Atlantic." *Collective Volume of Scientific Papers ICCAT* 45, no. 2 (1996): 358–67.

Ellis, R. *Deep Atlantic: Life, Death, and Exploration in the Abyss*. New York: Knopf, 1996.

———. *The Search for the Giant Squid*. New York: Lyons, 1998.

———. *Sea Dragons: Predators of the Prehistoric Oceans*. Lawrence: University Press of Kansas, 2003.

———. *The Empty Ocean*. Washington, DC: Island Press, 2003.

———. *Tuna: A Love Story*. New York: Knopf, 2008.

———. *Big Fish*. New York: Abrams, 2009.

Ellis, R., and J. McCosker. *Great White Shark*. Stanford, CA: Stanford University Press, 1991.

Emmanuel, K. "Increasing Destructiveness of Tropical Cyclones over the Past 30 Years." *Nature* 436 (2005): 686–88.

Erwin, D. H. *Extinction*. Princeton, NJ: Princeton University Press, 2006.

———. *The Great Paleozoic Crisis: Life and Death in the Permian*. New York: Columbia University Press, 1993.

———. "The Mother of Mass Extinctions." *Scientific American* 273, no. 1 (1996): 72–78.

———. "The Permo-Triassic Extinction." *Nature* 367 (1994): 231–36.

Eschmeyer, W. N. "A Deepwater-Trawl Capture of Two Swordfish (*Xiphias gladius*) in the Gulf of Mexico." *Copeia* 1963, no. 3 (1963): 590.

Everhart, M. J. *Oceans of Kansas: A Natural History of the Western Interior Sea*. Bloomington: Indiana University Press, 2005.

Faiella, G. *Fishing in Bermuda*. Oxford: Macmillan Caribbean, 2003.

Farrington, S.K. *Bill, the Broadbill Swordfish*. New York: Coward-McCann, 1942.

———. *Fishing the Atlantic, Offshore and On*. New York: Coward-McCann, 1949.

———. *Fishing the Pacific, Offshore and On*. New York: Coward-McCann, 1953.

———. *Fishing with Hemingway and Glassell*. New York: McKay, 1971.

Ferrell, D. "All over the Place." *Marlin* 30, no. 7 (2011): 8.

Fierstine, H. "Analysis and New Records of Billfish (Teleostei: Perciformes: Istiophoridae) from Yorktown Formation, Early Pliocene of Eastern North Carolina at Lee Creek Mine." *Smithsonian Contributions to Paleobiology* 90 (2001): 21–69.

———. "An Atlantic Blue Marlin, *Makaira nigricans*, Impaled by Two Species of Billfishes." *Bulletin of Marine Science* 61, no. 2 (1997): 495–99.

———. "Fossil History of Billfishes (Xiphioidei)." *Bulletin of Marine Science* 79, no. 3 (2006): 433–53.

———. "A New *Aglyptorhynchus* (Perciformes: Scombroidei: ?Blochiidae) from the Late Oligocene of Oregon." *Journal of Vertebrate Paleontology* 21, no. 1 (2001): 24–33.

———. "A new *Aglyptorhynchus* (Perciformes: Scombroidei) from the Lincoln Creek Formation (Late Oligocene, Washington, U.S.A.)" *Journal of Vertebrate Paleontology* 25, no. 2 (2005): 288–99.

———. "A New Marlin, *Makaira panamensis*, from the Late Miocene of Panama." *Copeia* 1978, no. 1 (1978): 1–11.

———. "A New Species of Xiphiorhynchid billfish (Perciformes; Scombroidei) from the Austrian Alps (Early Oligocene)." *Journal of Vertebrate Paleontology* 22, no. 3 (2002): 53A.

——. "*Makaira* sp., cf *M. nigricans Lacèpéde*, 1802 (Teleostei: Perciformes: Istiophoridae) from the Eastover Formation, Late Miocene, Virginia, and a Reexamination of *Istiophorus calvertensis* Berry, 1917." *Journal of Vertebrate Paleontology* 18, no. 1 (1998): 30–42.
——. "The Paleontology of Billfish—the State of the Art." In *Proceedings of the International Billfish Symposium, Kailua-Kona, Hawaii, 9–12 August, 1972*, edited by R. S. Shomura and F. Williams, 34–44. NOAA Technical Report NMSF SSRF 675. Seattle: National Marine Fisheries Service, 1974.
——. "A Paleontological Review of Three Billfish Families (Istiophoridae, Xiphiidae and Xiphiorhynchidae)." In *Planning the Future of Billfishes*. Pt. 2: *Contributed Papers*, edited by R. H. Stroud, 11–19. International Billfish Symposium, Kailua-Kona, Hawaii, August 1–5, 1988. Savannah, GA: National Coalition for Marine Conservation, 1990.
——. Review of *Pacific Marlins: Anatomy and Physiology* by Peter S. Davie. *Copeia* 1991, no. 4 (1991): 1160–61.
Fierstine, H., and S. P. Applegate. "Billfish Remains from Southern California with Remarks on the Importance of the Predentary Bone." *Bulletin of the Southern California Academy of Sciences* 67 (1968): 29–39.
——. "*Xiphiorhynchus kimbalocki*, a New Billfish from the Eocene of Mississippi with Remarks about the Systematics of Xiphioid Fishes." *Bulletin of the Southern California Academy of Sciences* 73, no. 1 (1974): 14–22.
Fierstine, H., S. P. Applegate, G. Gonzáles-Barba, T. Schwennicke, and L. Espinosa-Arrubarrena. "A Fossil Blue Marlin, *Makaira nigricans* Lacèpéde) from the Middle Faces of the Trinidad Formation (Upper Miocene to Upper Pliocene), San José del Cabo Basin, Baja California Sur, Mexico." *Bulletin of the Southern California Academy of Sciences* 100, no. 2 (2001): 59–73.
Fierstine, H., G. M. Cailliet, and J. M. Neer. "Shortfin Mako, *Isurus oxyrhinchus*, Impaled by Blue Marlin, *Makaira nigricans* (Teleostei: Istiophoridae)." *Bulletin of the Southern California Academy of Sciences* 96, no. 3 (1997): 117–21.
Fierstine, H., and O. Crimmen. "Two Erroneous, Commonly Cited Examples of 'Swordfish' Piercing Wooden Ships." *Copeia* 1996, no. 2 (1996): 472–75.
Fierstine, H., and K. A. Monsch. "Redescription and Phylogenetic Relationships of the Family Blochiidae (Perciformes: Scombroidei), Middle Eocene, Monte Bolca, Italy." *Miscellanea Paleontologica, Studi e Ricerche sui Giacimenti Terziari di Bolca (Museo Civico di Storia Naturale di Verona)* 9 (2002): 121–63.
Fierstine, H., and J. E. Starnes. "*Xiphiorhynchus* CF *X. eocanenicus* (Woodward, 1901), (Scombridaei: Xiphiidae: Xiphiorhnchinae) from the Middle

Eocene of Mississippi, the First Transatlantic Distribution of a Species of *Xiphiorhynchus*." *Journal of Vertebrate Paleontology* 25, no. 2 (2005): 280–87.

Fierstine, H., and N. Voigt. "Use of Rostral Characters for Identifying Adult Billfishes (Teleostei: Perciformes: Istiophoridae and Xiphiidae)." *Copeia* 1996, no. 1 (1996): 148–61.

Fierstine, H., and R. Weems. "A Fine Catch of Billfish from the Oligocene of South Carolina." Abstract. *Journal of Vertebrate Paleontology* 24, no. S3 (2004): 57a.

Fierstine, H., and B. J. Welton. "A Black Marlin, *Makaira indica*, from the Early Pleistocene of the Philippines and the Zoogeography of Istiophorid Billfishes." *Bulletin of Marine Science* 33, no. 3 (1983): 718–28.

———. "A Late Miocene Marlin, *Makaira* sp. (Perciformes, Osteichthyes) from San Diego County, California, U.S.A." *Tertiary Research* 10, no. 1 (1988): 13–19.

Fitch, J. E., and R. J. Lavenberg. *Marine Food and Game Fishes of California*. Berkeley: University of California Press, 1971.

Fitzgerald, G. *The Decline of the Cape Breton Swordfish Fishery: An Exploration of the Past and Recommendations for the Future of the Nova Scotia Fishery*. Marine Issues Committee Special Publication 6. Halifax: Ecology Action Center, 2000.

Fogt, J. "The Billfish Biologist." *Marlin*, July 2006, 82–87.

———. "Sailfish Evolution." *Marlin*, March 2005, 50–56.

Frazier, J. G., H. L. Fierstine, S. C. Beavers, F. Achaval, H. Suganuma, R. L. Pitman, Y. Yamaguchi, and C. M. Prigioni. "Impalement of Marine Turtles (Reptilia, Chelonia: Cheloniidae and Dermochelyidae) by Billfishes (Osteichthyes, Perciformes, Istiophoridae and Xiphiidae)." *Environmental Biology of Fishes* 39 (1994):85–96.

Friedman, M. "Ecomorphological Selectivity among Marine Teleost Fishes during the End-Cretaceous Extinction." *Proceedings of the National Academy of Sciences* 106, no. 3 (2009): 5218–23.

Fritsches, K. A., R. W. Brill, and E. J. Warrant. "Warm Eyes Provide Superior Vision in Swordfishes." *Current Biology* 15 (2005): 55–58.

Fritsches, K. A., N. J. Marshall, and E. J. Warrant. "Retinal Specializations in the Blue Marlin: Eyes Designed for Sensitivity at Low Light Levels." *Marine and Freshwater Research* 54, no. 4 (2003): 333–41.

Fritsches, K. A., J. C. Partridge, J. D. Pettigrew, and N. J. Marshall. "Colour Vision in Billfish." *Philosophical Transactions of the Royal Society of London* B 355 (2000):1253–56.

Fritsches, K. A. and E. Warrant. "Do Tuna and Billfish See Colors?" *Pelagic Fisheries Research Program Newsletter* 9, no. 1 (2004): 1–4.

Garcia-Cortés, B., J. Mejuto, and M. Quintans. "Summary of Swordfish (*Xiphias gladius*) Recaptures Carried out by the Spanish Surface Longline Fleet in the Atlantic Ocean: 1984–2002." *Collective Volume of Scientific Papers ICCAT* 55, no. 4 (2003): 1476–84.

Gibson, C. D. *The Broadbill Swordfishery in the Northwestern Atlantic*. Camden, ME: Ensign, 1998.

———. "A History of the Swordfishery in the Northwestern Atlantic." *American Neptune* 41, no. 1 (1981): 36–65.

Goadby, P. *Big Fish and Blue Water: Gamefishing in the Pacific*. Sydney: Angus & Robertson, 1975.

———. *Billfishing: The Quest for Marlin, Swordfish, Spearfish and Sailfish*. Camden, ME: International Marine, 1996.

Goode, G. B. "The Swordfish Fishery." In *The Fisheries and Fishing Industry of the United States*, G. B. Goode, 315–26. Washington: Government Printing Office, 1887.

Goodyear, C. P. "Blue Marlin Mean Length: Simulated Response to Increasing Fishing Mortality." *Marine and Freshwater Research* 54, no. 4 (2003): 401–8.

Gordon, B. L. "Sword-Bearer of the Seas." *Sea Frontiers* 25, no. 6 (1979): 357–63.

Gordon, M. "Swordfish Lore." *Natural History* 36 (1935): 319–26.

Gosline, A. "Simple Changes Could Save Swordfish." *New Scientist*, October 3, 2004. http://www.newscientist.com/article/dn6455-simple-changes-could-save-swordfish.html.

Govender, A., R. van der Elst, and N. James. *Swordfish: Global Lessons*. WWF South Africa, 2003. http://www.oceandocs.net/bitstream/1834/921/1/Swordfish.pdf.

Govoni, J. J., E. H. Laban, and J. A. Hare. "The Early Life History of Swordfish (*Xiphias gladius*) in the Western North Atlantic." *Fishery Bulletin* 101, no. 4 (2003): 778–89.

Graves, J. E. "Billfish Science." *Marlin* 30, no. 7 (2001): 58–62.

———. "Molecular Insights into the Population Structures of Cosmopolitan Marine Fishes." *Journal of Heredity* 89, no. 5 (1998): 427–37.

Graves, J. E., D. W. Kerstetter, B. E. Luckhurst, and E. D. Prince. "Habitat Preference of Istiophorid Billfishes in the Western North Atlantic: Applicability of Archival Tag Data to Habitat-Based Stock Assessment Methodologies." *Collective Volume of Scientific Papers ICCAT* 55, no. 2 (2003): 594–602.

Graves, J. E. and J. R. McDowell. "Genetic Analysis of White Marlin (*Tetrapterus albidus*) Stock Structure." *Bulletin of Marine Science* 79, no. 3 (2006): 469–82.

———. "Stock Structure of the World's Istiophorid Billfishes: A Genetic Perspective." *Marine and Freshwater Research* 54, no. 4 (2003): 287–98.

Greenlaw, L. *The Hungry Ocean: A Swordboat Captain's Journey*. New York: Hyperion, 1999.

Gregory, W. K., and G. M. Conrad. "The Comparative Osteology of the Swordfish (*Xiphias*) and the Sailfish (*Istiophorus*)." *American Museum Novitates* 952 (1937): 1–25.

Grey. R. C. *Adventures of a Deep-Sea Angler*. New York: Harper & Brothers, 1930. Reprint, Lanham, MD: Derrydale Press, 2002.

Grey, Z. *An American Angler in Australia*. New York: Harper & Brothers, 1937. Reprint, Mattituck, NY: American Reprint Co., 1996.

———. *Tales of the Angler's Eldorado, New Zealand*. New York: Harper & Brothers, 1926. Reprinted as *Angler's Eldorado: Zane Grey in New Zealand*. New York: Walter J. Black, 1982.

———. *Tales of Fishes*. New York: Harper & Brothers, 1919. Reprint, Lyon, MS: Derrydale Press, 1990.

———. *Tales of Fishing Virgin Seas*. New York: Harper & Brothers, 1925.

———. *Tales of Swordfish and Tuna*. New York: Harper & Brothers, 1927.

———. *Tales of Tahitian Waters*. New York: Harper & Brothers, 1931. Reprint, Lyon, MS: Derrydale Press, 1990.

Gudger, E. W. "The Alleged Pugnacity of the Swordfish and the Spearfishes as Shown by Their Attacks on Vessels." *Memoirs of the Royal Asiatic Society of Bengal* 12, no. 2 (1940): 215–315.

Guerra, A., F. Simon, and A. F. Gonzalez. "Cephalopods in the Diet of the Swordfish *Xiphias gladius*, from the Northeastern Atlantic Ocean." In *Recent Advances in Cephalopod Fisheries Biology*, edited by T. Okutani, R. K. O'Dor, and T. Kubodera, 159–64. Tokyo: Tokai University Press, 1993.

Günther, A. C. L. G. *An Introduction to the Study of Fishes*. Edinburgh: Adam and Charles Black, 1880.

Habron, G. B., P. M. Mace, S. Koplin, and G. P. Scott. "United States Imports of Swordfish (1974–June 1994)." *Collective Volume of Scientific Papers ICCAT* 44, no. 3 (1995): 174–78.

Hallam, A. "The Case for Sea-Level Change as a Dominant Causal Factor in Mass Extinction of Marine Invertebrates." *Philosophical Transactions of the Royal Society of London* B 325 (1989): 437–55.

Hallam, A., and P. B. Wignall. *Mass Extinctions and Their Aftermath*. Oxford: Oxford University Press, 1997.

Hamabe, M., C. Hamuro, and M. Ogura. *Squid Jigging from Small Boats*. FAO Fishing Manuals. Farnham, Surrey, England: Published by arrangement with the Food and Agriculture Organization of the United Nations by Fishing News Books, 1982.

Hanlon, R. T., and J. B. Messenger. *Cephalopod Behaviour*. Cambridge: Cambridge University Press, 1996.

Hansen, J. "Defusing the Global Warming Time Bomb." *Scientific American* 290, no. 3 (2004): 68–77.

Harvey, G. C. McN. "An Historical Review of Recreational and Artisanal Fisheries for Billfish in Jamaica, 1796–1988." *Collective Volume of Scientific Papers ICCAT* 30, no. 2 (1989): 440–50.

———. *Portraits from the Deep*. Winter Park, FL: World Publications, 2002.

Hayden, T. "Empty Oceans: Why the World's Seafood Supply Is Disappearing." *U.S. News and World Report* 134, no. 20 (2003): 38–45.

Heide-Jørgensen, M. P. Narwhal. In *Encyclopedia of Marine Mammals*, edited by W. F. Perrin, B. Würsig, and J. G. M. Thewissen, 754–58. Boston: Academic Press, 2009.

Heilner, V. C. *Salt Water Fishing*. New York: Knopf, 1953.

Hemingway, E. *Death in the Afternoon*. New York: Scribner's, 1932.

———. *A Farewell to Arms*. New York: Scribner's, 1929.

———. *Green Hills of Africa*. New York: Scribner's, 1935.

———. *Islands in the Stream*. New York: Scribner's, 1970.

———. *The Old Man and the Sea*. New York: Scribner's, 1952.

Hendrickson, P. *Hemingway's Boat*. New York: Knopf, 2011.

Herald, E.S. *Living Fishes of the World*. New York: Doubleday, 1961.

Hernandez-Garcia, V. "The Diet of the Swordfish *Xiphias gladius* Linnaeus 1758, in the Central East Atlantic, with Emphasis on the Role of Cephalopods." *Fishery Bulletin* 93, no. 2 (1995): 403–11.

Hernández-Herrera, A., E. Morales-Borórquez, M. A. Cisneros-Mata, M. O. Nevárez-Martínez, and G. I. Rivera-Parra. Management Strategy for the Giant Squid (*Dosidicus gigas*) Fishery in the Gulf of California, Mexico. *California Cooperative Oceanic Fisheries Investigations Report* 39 (1998): 212–18.

Hess, S. C., and R. B. Toll. "Methodology for Specific Diagnosis of Cephalopod Remains in Stomach Contents of Predators with Reference to the Broadbill Swordfish, *Xiphias gladius*." *Journal of Shellfish Research* 1, no. 2 (1981): 161–70.

Hinton, M. G. "Status of Swordfish Stocks in the Eastern Pacific Ocean Estimated Using Data from Japanese Tuna Longline Fisheries." *Marine and Freshwater Research* 54, no. 4 (2003): 393–99.

Hoey, J. J., and A. Bertolino. "Review of the U.S. Fishery for Swordfish, 1978 to 1986." *Collective Volume of Scientific Papers ICCAT* 27 (1988): 256–66.

Hoey, J. J., and J. Casey. "Review of the U.S. Fishery for Swordfish, 1960 to 1977." *Collective Volume of Scientific Papers ICCAT* 27 (1988): 267–82.

Hoey, J. J., and J. Mejuto. "Swordfish Size Composition Data from Spanish and United States North Atlantic Longline Fisheries." *Collective Volume of Scientific Papers ICCAT* 35, no. 2 (1991): 415–28.

Hoey, J. J., J. Mejuto, S. Iglesias, and R. Cosner. "A Comparative Study of the United States and Spanish Longline Fleets Targeting Swordfish in the Atlantic Ocean North of 40° Latitude." *Collective Volume of Scientific Papers ICCAT* 27 (1988): 230–39.

Holland, K. N. "A Perspective on Billfish Biological Research and Recommendations for the Future." *Marine and Freshwater Research* 54, no. 4 (2003): 343–47.

Hosaka, E. Y. *Sport Fishing in Hawaii*. Honolulu: Bond's, 1944.

Hoyos, C. D., P. A. Agudelo, P. J. Webster, and J. A. Curry. "Deconvolution of the Factors Contributing to the Increase in Global Hurricane Intensity." *Science* 312 (2006): 94–97.

Hsieh, C.-H., C. S. Reiss, J. R. Hunter, J. R. Beddington, R. M. May, and G. Sugihara. "Fishing Elevates Variability in the Abundance of Exploited Species." *Nature* 443 (2006): 859–62.

Hyerdahl, T. *The Kon-Tiki Expedition*. London: Allen & Unwin, 1950.

Ibáñez, C. M., C. González, and L. Cubillos. "Dieta del Pez Espada *Xiphias gladius* Linnaeus, 1758, en Aguas Oceánicas de Chile Central en Invierno de 2003." *Investigaciones Marinos de Valparaiso* 32, no. 2 (2004): 113–20.

International Convention for the Conservation of Atlantic Tunas. "1998 Swordfish—Detailed Report." *Collective Volume of Scientific Papers ICCAT* 49, no. 1 (1999): 175–210.

IUCN. 2011. *2011 IUCN Red List of Threatened Species*. www.iucnredlist.org.

Jackson, J. B. C., and K. G. Johnson. "Measuring Past Biodiversity." *Science* 293 (2001): 2401–3.

Jackson, J. B. C., M. X. Kirby, W. H. Berger, K. A. Bjorndal, L. W. Botsford, B. J. Bourque, R. H. Bradbury, et al. "Historical Overfishing and the Recent Collapse of Coastal Ecosystems." *Science* 293 (2001): 629–38.

Jin, Y. G., Y. Wang, W. Wang, Q. H. Shang, C. Q. Cao, and D. H. Erwin. "Pattern of Marine Mass Extinction near the Permian-Triassic Boundary in South China." *Science* 289 (2000): 432–36.

Johnson, G. D., and A. C. Gill. "Perches and Their Allies." In *Encyclopedia of*

Fishes, edited by J. R. Paxton and W. N. Eschmeyer, 181–96. San Diego, CA: Academic Press, 1995.

Jones, E. C. "*Isistius brasiliensis*, a Squaloid Shark, the Probable Cause of Crater Wounds on Fishes and Cetaceans." *Fishery Bulletin* 69, no. 4 (1971): 791–98.

Jonsgård, Å. "New Find of Sword from Swordfish (*Xiphias gladius*) in Blue Whale (*Balaenoptera musculus*)." *Norsk Hvalfangst-Tidende* 48, no. 7 (1959): 352–60.

———. "Three Finds of Swords from Swordfish (*Xiphias gladius*) in Antarctic Fin Whales (*Balaenoptera physalus* [L])." *Norsk Hvalfangst-Tidende* 51, no. 7 (1962): 287–91.

Jordan, D. S., and B. W. Evermann. *American Food and Game Fishes*. New York: Doubleday, Page & Co., 1902.

———. *The Fishes of North and Middle America*. 4 pts. Bulletin of the US National Museum, no. 47. Washington: Government Printing Office, 1900.

———. "A Review of the Giant Mackerel-Like Fishes, Tunnies, Spearfishes, and Swordfishes." *Occasional Papers of the California Academy of Sciences* 12 (1926): 1–113.

Joseph, J., and J. W. Greenough. *International Management of Tuna, Porpoise, and Billfish*. Seattle: University of Washington Press, 1979.

Joseph, J., W. Klawe, and P. Murphy. *Tuna and Billfish: Fish without a Country*. La Jolla, CA: Inter-American Tropical Tuna Commission, 1988.

Josselyn, J. *An Account of Two Voyages to New England, London*. 1675. Reprint, Boston: Wm. Veazie, 1865.

Julian, S. "The Mighty Sword? As the Boycotted Fish Migrate, the Market Remains in a Muddle." *Boston Globe*, September 30, 1998.

Junger, S. *The Perfect Storm*. New York: HarperPerennial, 1999.

Kerr, R. A. "Sea Change in the Atlantic." *Science* 303 (2004): 35.

———. "A Worrying Trend of Less Ice, Higher Seas." *Science* 311 (2006): 1698–1701.

Kingsley, J. S. "Food Habits of Swordfish." *Science* 56 (1922): 225–26.

Kiraly, S. J., J. A. Moore, and P. H. Jasinski. "Deepwater and Other Sharks of the U.S., Atlantic Ocean Exclusive Economic Zone." *Marine Fisheries Review* 65, no. 4 (2003): 1–65.

Klimley, A. P., J. E. Richert, and S. J. Jorgensen. "The Home of Blue Water Fish." *American Scientist* 93, no. 1 (2005): 42–49.

Knoll, A. H., R. K. Bambach, D. E. Canfield, and J. P. Grotzinger. "Comparative Earth History and Late Permian Mass Extinction." *Science* 273 (1996): 452–57.

Kotoulas, G., J. Mejuto, G. Tserpes, B. Garcia-Cortés, P. Peristeraki, J. M. de la Serna, and A. Magoulas. "DNA Microsatellite Markers in Service of

Swordfish Stock-Structure Analysis in the Atlantic and Mediterranean." *Collective Volume of Scientific Papers ICCAT* 55, no. 4 (2003): 1632–39.
Krumholz, L. A., and D. P. DeSylva. "Some Foods of Marlins near Bimini, Bahamas." *Bulletin of the American Museum of Natural History* 114, no. 5 (1958): 406–11.
Kume, S., and J. Joseph. "Size Composition and Sexual Maturity of Billfishes Caught by the Japanese Longline Fishery in the Pacific Ocean East of 130°." *Bulletin of the Far Seas Research Laboratory* 2 (1969): 115–62.
LaMonte, F. *Giant Fishes of the Open Sea*. New York: Holt, Rinehart & Winston, 1965.
———. *Marine Game Fishes of the World*. New York: Doubleday, 1952.
———. *North American Game Fishes*. New York: Doubleday, 1945.
Lee, M., ed. *Seafood Lover's Almanac*. Islip, NY: National Audubon Society, 2000.
Leech, M. "ICCAT: A Paper Tiger." *Marlin*, March 2012, 32–34.
———. "New Regs for Anglers." *Marlin*, July 2006, 36–38.
———. "Resurrection." *Marlin*, February 2006, 80–84.
———. "Status Unknown: Keeping Track of Atlantic Sailfish Populations." *Marlin*, October 2005, 32–34.
———. "Swordfish Recovery." *Marlin*, May 2005, 36–38.
Levine, C. "Marlin Punch: Black Marlin Injures Teen in Panama." *Marlin*, October 2005, 60–63.
Levitus, S., J. I. Antonov, T. P. Boyer, and C. Stephens. "Warming of the World Ocean." *Science* 287 (2000): 2225–29.
Lineaweaver, T. H., and R. H. Backus. *The Natural History of Sharks*. New York: Anchor Doubleday, 1973.
Low, F. H. *Fishing Is for Me*. New York: William Morrow, 1963.
Lowry, M., and J. Murphy. "Monitoring the Recreational Gamefish Fishery off South-Eastern Australia." *Marine and Freshwater Research* 54, no. 4 (2003): 425–34.
Luckhurst, B. E. "Historical Development of Recreational Billfishing in Bermuda and the Significance of Catches of Large Blue Marlin (*Makaira nigricans*)." *Marine and Freshwater Research* 54, no. 4 (2003): 459–62.
Major, P. F. "An Aggressive Encounter between a Pod of Whales and Billfish." *Scientific Reports of the Whales Research Institute* 31 (1979): 95–96.
———. "Combat on the High Seas." *Sea Frontiers* 27 (1981): 280–86.
Maksimov, V. P. "Swordfish Attack on a Shark." *Problems of Ichthyology* 8, no. 1 (1968): 756.
Markaida, U. "Cephalopods in the Diet of Swordfish (*Xiphias gladius*) Caught

off the West Coast of Baja California, Mexico." *Pacific Science* 59, no. 1 (2005): 25–41.
Markaida, U., C. Quiñónes-Velasquéz, and O. Sosa-Nishizaki. "Age, Growth and Maturation of the Jumbo Squid *Dosidicus gigas* (Cephalopoda: Ommastrephidae) from the Gulf of California, Mexico." *Fisheries Research* 66 (2004): 31–47.
Markaida, U., J. J. C. Rosenthal, and W. F. Gilly. "Tagging Studies on the Jumbo Squid (*Dosidicus gigas*) in the Gulf of California, Mexico." *Fishery Bulletin* 103 (2005): 219–26.
Markaida, U., and O. Sosa-Nishizaki. "Food and Feeding Habits of Swordfish *Xiphias gladius* L., off Western Baja California." In *Biology and Fisheries of Swordfish,* Xiphias gladius*: Papers from the International Symposium on Pacific Swordfish, Ensenda, Mexico, 11–14 December 1994*, edited by I. Barrett, O. Sosa-Nishizaki and N. Bartoo, 245–59. NOAA Technical Report, no. 142. Seattle: US Department of Commerce, National Oceanic and Atmospheric Administration, National Marine Fisheries Service, Scientific Publications Office, 1998.
Marron, E. *Albacora: The Search for the Giant Broadbill.* New York: Random House, 1957.
Marshall, N. B. *The Life of Fishes.* Winter Park, FL: World Publications, 1966.
Masters, R. "Real Sea Monsters." *Monterey County Weekly*, March 10, 2005, 1–6.
Mather, C. O. *Billfish: Marlin, Broadbill, Sailfish.* Sydney, BC: Saltaire, 1976.
Mather, J. A., and R. K. O'Dor. "Spatial Organization of Schools of the Squid *Illex illecebrosus.*" *Marine Behaviour and Physiology* 10 (1984): 259–71.
Matsen, B. *Deep Sea Fishing: The Lure of Big Game Fish.* San Diego, CA: Thunder Bay Press, 1990.
Matthiessen, P. *Men's Lives: The Surfmen and Baymen of the South Fork.* New York: Random House, 1986.
Mattiucci, S., V. Farina, A. Garcia, M. N. Santos, L. Marinello, and G. Nascenti. "Metazoan Parasitic Infections of Swordfish (*Xiphias gladius* L., 1758) from the Mediterranean Sea and Atlantic Gibraltar Waters: Implications for Stock Assessment." *Collective Volume of Scientific Papers ICCAT* 58, no. 4 (2005): 1470–82.
McCarty, J. P. "Ecological Consequences of Recent Climate Change." *Conservation Biology* 15, no. 2 (2001): 320–31.
McDowell, J. R., and J. E. Graves. "Population Structure of Striped Marlin (*Kajikia audax*) in the Pacific Ocean Based on Analysis of Microsatellite and Mitochondrial DNA." *Canadian Journal of Fisheries and Aquatic Sciences* 65 (2008): 1307–20.

McGowan, C. "Differential Development of the Rostrum and Mandible of the Swordfish (*Xiphias gladius*) during Ontogeny and Its Possible Functional Significance." *Canadian Journal of Zoology* 66 (1988): 496–503.
———. "A Putative Ancestor for the Swordfish-Like Ichthyosaur *Eurhinosaurus*." *Nature* 322, no. 6078 (1986): 454–56.
Mejuto, J., U, Autón, and M. Quintans. "Visual Acuity and Olfactory Sensitivity in the Swordfish (*Xiphias gladius*) for the Detection of Prey during Field Experiments Using the Surface Longline Gear with Different Bait Types." *Collective Volume of Scientific Papers ICCAT* 55, no. 4 (2005): 1501–10.
Mejuto, J., and B. Garcia-Cortés. "A Description of a Possible Spawning Area of the Swordfish (*Xiphias gladius*) in the Tropical Northwest Atlantic." *Collective Volume of Scientific Papers ICCAT* 55, no. 4 (2003): 1449–58.
Mejuto, J., B. Garcia-Cortés, J. M. de la Serna, and A. Ramos-Cartelle. "An Overview of the Activity of the Spanish Surface Longline Fleet Catching Swordfish (*Xiphias gladius*) during the Year 2002, with Special Reference to the Atlantic Ocean." *Collective Volume of Scientific Papers ICCAT* 58, no. 4 (2005): 1495–1500.
Meltzoff, S. "Like a Neon Shadow in the Sea." *Sports Illustrated* 47, no. 4 (1977): 22–29.
Meneses de Lima, J. H., J. E. Kotas, and C. F. Lin. "A Historical Review of the Brazilian Longline Fishery and Catch of Swordfish (1972–1997)." *Collective Volume of Scientific Papers ICCAT* 51, no. 1 (2000): 1329–57.
Migdalski, E. C. *Angler's Guide to the Salt Water Game Fishes, Atlantic and Pacific*. New York: Ronald Press, 1958.
Migdalski, E. C., and G. S. Fichter. *The Fresh and Salt Water Fishes of the World*. New York: Knopf, 1976.
Miles, J. "Rider of the Purple Prose." Review of *Zane Grey: His Life, His Adventures, His Women*, by Thomas H. Pauly. *New York Times Book Review*, January 1, 2005, 8–9.
Milius, S. "That's One Weird Tooth (and Other Bulletins on the Elusive Narwhal)." *Science News* 169, no. 12 (2006): 186–88.
Mitchell Hedges, F. A. *Battles with Giant Fish*. London: Duckworth, 1923.
Monsch, K. A., H. L. Fierstine, and R. E. Weems. "Taxonomic Revisions and Stratigraphic Provenance of '*Histiophorus rotundus*' Woodward 1901 (Teleostei, Perciformes)." *Journal of Vertebrate Paleontology* 25, no. 2 (2005): 274–79.
Montocchio, M. "White Marlin or Spearfish?" *Marlin* 31, no. 1 (2012): 44–49.
Moreira, F. "Food of the Swordfish *Xiphias gladius*, Linnaeus 1758, off the Portuguese Coast." *Journal of Fish Biology* 36 (1990): 623–24.

Morrow, J. E. "A Striped Marlin (*Makaira mitsukurii*) without a Spear." *Copeia* 1951 (1951): 303–4.

Mossman, S. "Light Tackle, Fat Stripers." *Marlin*, November 2005, 38–43.

Mowbray, L. L. "Certain Citizens of Warm Seas." In *The Book of Fishes*, edited by J. O. La Gorce, 196–267. Washington, DC: National Geographic Society, 1952.

Moyle, P. B., and J. J. Cech. *Fishes: An Introduction to Ichthyology*. Upper Saddle River, NJ: Prentice-Hall, 2004.

Mudge, B. F. "Rare Forms of Fish in Kansas." *Transactions of the Kansas Academy of Science* 3 (1874): 121–22.

Mundus, F., and W. L. Wisner. *Sportfishing for Sharks*. New York: Macmillan, 1971.

Muñoz-Chápuli, R., J. C. Rey Salgado, and J. M. de la Serna. "Bio-geography of *Isistius brasiliensis* in the North-Eastern Atlantic, Inferred from Crater Wounds on Swordfish." *Journal of the Marine Biological Association of the U.K.* 68, no. 2 (1988): 315–21.

Myers, G. J., P. W. Davidson, C. Cox, C. F. Shamlaye, D. Palumbo, E. Cerniciari, J. Sloan-Reeves, et al. "Prenatal Methylmercury Exposure from Ocean Fish Consumption in the Seychelles Child Development Study." *Lancet* 361 (2003): 1686–92.

Myers, R. A., and B. Worm. "Rapid Worldwide Depletion of Predatory Fish Communities." *Nature* 423 (2003): 280–83.

Nagle, M. "Prehistoric Swordfishing." *UMaine Today* 5, no. 2 (2005): 15–18.

Nakamura, I. *Billfishes of the World: An Annotated and Illustrated Catalogue of Marlins, Sailfishes, Spearfishes and Swordfishes Known to Date*. Vol. 5 of *FAO Species Catalogue*. FAO Fisheries Synopsis, no. 125. Rome: United Nations Development Programme Food and Agriculture Organization, 1985.

National Oceanic and Atmospheric Administration. *Draft Amendment 1 to the Fishery Management Plan for Atlantic Swordfish, including an Environmental Regulatory Impact Review*. Washington: Government Printing Office, 1997.

Nichols, J. T., and F. R. LaMonte. "How Many Marlins Are There?" *Natural History* 36 (1935): 327–30.

———. "Notes on Swordfish at Cape Breton, Nova Scotia." *American Museum Novitates* 901 (1937): 1–7.

———. "The Tahitian Black Marlin, or Silver Marlin Swordfish." *American Museum Novitates* 807 (1935): 1–2.

Nigmatullin, Ch. M., K. N. Nesis, and A. I. Arkhipkin. "A Review of the Biology of the Jumbo Squid *Dosidicus gigas* (Cephalopoda: Ommastrephidae)." *Fisheries Research* 54 (2001): 9–19.

Nigrelli, R. F. "Parasites of the Swordfish, *Xiphias gladius* Linnaeus." *American Museum Novitates* 996 (1938): 1–16.

Norman, J. R., and F. C. Fraser. *Giant Fishes, Whales and Dolphins*. New York: W. W. Norton, 1938.

Norman, J. R., and P. H. Greenwood. *A History of Fishes*. New York: Hill and Wang.

Northridge, S. P. 1991. *Driftnet Fisheries and Their Impacts on Non-Target Species: A Worldwide Review*. FAO Fisheries Technical Paper, no. 320. Rome: Food and Agriculture Organization of the United Nations, 1963.

Ohsumi, S. "Find of Marlin Spear from the Antarctic Minke Whales." *Scientific Reports of the Whales Research Institute* 25 (1973): 237–39.

Oken, E., K. P. Kleinman, W. E. Berland, S. R. Simon, J. W. Rich-Edwards, and M. W. Gillman. "Decline in Fish Consumption among Pregnant Women after a National Mercury Advisory." *Obstetrics and Gynecology* 102 (2003): 346–51.

Oppian. *Halieutica*. In *Oppian, Colluthus, Tryphiodorus*. Trans. A. W. Mair. Loeb Classical Library, no. 219. Cambridge MA: Harvard University Press, 1928.

Ortiz, M. "Standardized Catch Rates by Sex and Age for Swordfish (*Xiphias gladius*) from the U.S. Longline Fleet 1981–2003." *Collective Volume of Scientific Papers ICCAT* 58, no. 4 (2005): 1446–65.

Ortiz, M., E. D. Prince, J. E. Serafy, D. B. Holts, K. B. Davy, J. G. Pepperell, M. B. Lowry, and J. C. Holdsworth. "Global Overview of the Major Constituent-Based Billfish-Tagging Programs and Their Results since 1954." *Marine and Freshwater Research* 54, no. 4 (2003): 489–507.

Ovchinnikov, V. V. *Swordfishes and Billfishes in the Atlantic Ocean: Ecology and Functional Morphology*. Trans. H. Mills. Jerusalem: Israel Program for Scientific Translation, 1971.

Paul, L. M. B. *High Seas Driftnetting: The Plunder of the Global Commons*. Kailua, HI: Earthtrust, 1994.

Pauly, D., V. Christensen, J. Dalsgaard, R. Froese, and F. Torres. "Fishing down Marine Food Webs." *Science* 279 (1998): 860–63.

Pauly, D., V. Christensen, R. Froese, and M. L. Palomares. "Fishing down Aquatic Food Webs." *American Scientist* 88, no. 1 (2000): 46–51.

Pauly, D., and J. Maclean. *In a Perfect Ocean: The State of Fisheries and Ecosystems in the North Atlantic Ocean*. Washington, DC: Island Press, 2002.

Peel, E., R. Nelson, and C. P. Goodyear. "Managing Atlantic Marlin as Bycatch under ICCAT. The Fork in the Road: Recovery or Collapse." *Marine and Freshwater Research* 54, no. 4 (2003): 575–84.

Peri, A. "Contaminated Swordfish Found in California Grocery Stores: Mercury Levels in Swordfish Dangerously High Says New Data." Sea Turtle Restoration Project. Press release, September 28, 2004. http://seaturtles.org/article.php?id=739.

Peristeraki, P., N. Kypraios, G. Lazarakis, and G. Tserpes. "By-Catches and Discards of the Greek Swordfish Fishery." *Collective Volume of Scientific Papers ICCAT* 62, no. 4 (2008): 1052–57.

Perkins, S. "The Big Fish That Went Away . . ." *Science News* 166 (2004): 334.

Piccard, J. *The Sun beneath the Sea*. New York: Scribner's, 1971.

Pierce, W. G. *Going Fishing: The Story of the Deep-Sea Fishermen of New England*. Camden, ME: International Marine, 1989.

Pliny. *Natural History*. 10 vols. Trans. H. Rackham. Loeb Classical Library. Cambridge, MA: Harvard University Press, 1938–63.

Prince, E. D., J. Luo, C. P. Goodyear, J. P. Hoolihan, D. Snodgrass, E. S. Orbesen, J. E. Serafy, M. Ortiz, and M. J. Schirripa. "Ocean Scale Hypoxia-Based Habitat Compression of Atlantic Istiophorid Billfishes." *Fisheries Oceanography* 19, no. 6 (2010): 448–62.

Prince, E. D., M. Ortiz, and A. Venizelos. "A Comparison of Circle Hook and 'J' Hook Performance in Recreational Catch-and-Release Fisheries for Billfish." In *Catch and Release in Marine Recreational Fisheries*, edited by J. A. Lucy and A. L. Studholme, 66–79. American Fisheries Society, Symposium 30. Bethesda, MD: American Fisheries Society, 2002.

Radcliffe, W. *Fishing from the Earliest Times*. London: John Murray, 1921. Reprint, Chicago: Ares Publishers, 1974.

Raloff, J. "Hammered Saws." *Science News* 172, no. 6 (2007): 90–92.

Ramos-Cartelle, A., and J. Mejuto. "Interaction of the False Killer Whale (*Pseudorca Crassidens*) and Depredation on the Swordfish Catches of the Spanish Surface Longline Fleet in the Atlantic, Indian, and Pacific Oceans." *Collective Volume of Scientific Papers ICCAT* 62, no. 6 (2008): 1721–38.

Raup, D. M. "Extinction: Bad Genes or Bad Luck?" *New Scientist* 131, no. 1786 (1991): 46–49.

———. *Extinction: Bad Genes or Bad Luck?* New York: Norton, 1991.

Raven, H. C., and F. R. LaMonte. "Notes on the Alimentary Tract of the Swordfish (*Xiphias gladius*)." *American Museum Novitates* 902 (1937): 1–13.

Reeb, C. A., L. Arcangeli, and B. A. Block. "Structure and Migration Corridors in Pacific Populations of the Swordfish *Xiphias gladius*, as Inferred through Analyses of Mitochondrial DNA." *Marine Biology* 136, no. 6 (2000): 1123–31.

Reeves, R. R. and E. D. Mitchell. "The Whale behind the Tusk." *Natural History* 90, no. 8 (1981): 50–57.

Reiger, G. *Profiles in Saltwater Angling*. Upper Saddle River, NJ: Prentice-Hall, 1973.

———. "Where Have All the Marlin Gone?" *Field and Stream* 92, no. 11 (1988): 77–85.

Relini, O., L. F. Garibaldi, C. Cima, and G. Palandri. "Feeding of the Swordfish, the Bluefin and Other Pelagic Nekton in the Western Ligurian Sea." *Collective Volume of Scientific Papers ICCAT* 44, no. 1 (1995): 283–86.

Restrepo, V., E. D. Prince, G. P. Scott, and Y. Uozumi. "ICCAT Stock Assessments of Atlantic Billfish." *Marine and Freshwater Research* 54, no. 4 (2003): 361–67.

Retallack, G. J., R. M. H. Smith, and P. D. Ward. "Vertebrate Extinction across Permian-Triassic Boundary in Karoo Basin, South Africa." *Geological Society of America Bulletin* 115, no. 9 (2003): 1133–52.

Revkin, A. C. "Commercial Fishing Is Cited in Decline of Oceans' Big Fish." *New York Times*, May 14, 2003.

Ricciuti, E. R. *Killers of the Seas*. New York: Walker, 1973.

Richardson, S. "Warm Blood for Cold Water." *Discover* 15, no. 1 (1994): 42–43.

Rivkin, M. *Big Game Fishing Headquarters: A History of the IGFA*. Dania Beach, FL: IGFA Press, 2005.

Robbins, M. W. "The Catch." *Mother Jones* 31, no. 2 (2006): 49–53.

Robins, C. R., and D. P. de Sylva. "Description and Relationships of the Longbill Spearfish *Tetrapturus belone*, Based on Western North Atlantic Specimens." *Bulletin of Marine Sciences of the Gulf and Caribbean* 10, no. 4 (1960): 383–413.

———. "A New Western Atlantic Spearfish, *Tetrapturus pfluegeri*, with a Redescription of the Mediterranean Spearfish, *Tetrapturus belone*." *Bulletin of Marine Sciences of the Gulf and Caribbean* 13, no. 1 (1963): 84–122.

Roe, S., and M. Hawthorne. "How Safe Is Tuna?" *Chicago Tribune*, December 13, 2005.

Roper, C. F. E., M. J. Sweeney, and C. E. Nauen. *Cephalopods of the World: An Annotated and Illustrated Catalogue of Species of Interest to Fisheries*. Vol. 3 of *FAO Species Catalogue*. FAO Fisheries Synopsis, no. 125. Rome: United Nations Development Programme Food and Agriculture Organization, 1984.

Ryskin, G. "Methane-Driven Oceanic Eruptions and Mass Extinctions." *Geology* 31, no. 9 (2003): 741–44.

Safina, C. "Fish Market Mutiny." *New York Times*, April 14, 1998.

———. *Song for the Blue Ocean*. New York: Henry Holt, 1997.

———. "Song for the Swordfish." *Audubon* 100, no. 3 (1998): 58–69.

———. *Voyage of the Turtle*. New York: Henry Holt, 2006.

———. "The World's Imperiled Fish." *Scientific American* 273, no. 5 (1995): 46–53.

Safina, C., A. A. Rosenberg, R. A. Myers, T. J. Quinn, and J. S. Collie. "U.S. Ocean Fish Recovery: Staying the Course." *Science* 309 (2005): 707–8.

Saito, H., M. Takahashi, K. Yokara, and Y. Uozumi. "Recent Status of Blue and White Marlin Catches by the Japanese Longline Fishery in the Atlantic Ocean." *Collective Volume of Scientific Papers ICCAT* 56, no. 3 (2001): 365–70.

Salman, A. "The Role of Cephalopods in the Diet of Swordfish (*Xiphias gladias* Linnaeus 1758) in the Aegean Sea (Eastern Mediterranean)." *Bulletin of Marine Research* 74, no. 1 (2004): 21–29.

Samson, J. *Line Down! The Special World of Big-Game Fishing*. [New York]: Winchester, 1973.

Sanger, D. *Discovering Maine's Archaeological Heritage*. Augusta: Maine Historic Preservation, 1979.

Schlesinger, M. *Avoiding Dangerous Climate Change*. Cambridge: Cambridge University Press, 2006.

Schneider. V. P., and H. Fierstine. "Fossil Tuna Vertebrae Punctured by Istiophorid Billfishes." *Journal of Vertebrate Paleontology* 24, no. 1 (2004): 253–55.

"Science Examines Record Mako Shark." *Long Island Traveler—Watchman*, August 11, 1977, 28.

Scott, W. B., and S. N. Tibbo. "Food and Feeding Habits of the Swordfish, *Xiphias gladias*, in the Western North Atlantic." *Journal of the Fisheries Research Board of Canada* 25, no. 5 (1968): 903–19.

Serafy, J. E., R. K. Cowen, C. B. Paris, T. R. Capo, and S. A. Luthy. "Evidence of Blue Marlin, *Makaira Nigricans*, Spawning in the Vicinity of Exuma Sound, Bahamas." *Marine and Freshwater Research* 54, no. 4 (2003): 299–306.

Serafy, J. E., G. A. Diaz, E. D. Prince, E. S. Orbesen, and C. M. Legault. "Atlantic Blue Marlin, *Makaira Nigricans*, and White Marlin, *Tetrapturus albidus*, Bycatch of the Japanese Pelagic Longline Fishery, 1960–2000." *Marine Fisheries Review* 66, no. 2 (2004): 9–20.

Shelly, K. C. *Lynn Bogue Hunt: A Sporting Life*. Lanham, MD: Derrydale Press, 2003.

Shimose, T., K. Yokawa, H. Saito, and K. Tachihara, K. "Evidence for Use of the Bill by Blue Marlin, *Makaira nigricans*, during Feeding." *Ichthyological Research* 54, no. 4 (2007): 420–22.

Shivji, M. S., J. E. Magnussen, L. R. Beerkircher, G. Hinteregger, D. W. Lee, J. E. Serafy, and E. D. Prince. "Validity, Identification, and Distribution of the Roundscale Spearfish, *Tetrapterus georgii* (Teleostei: Istiophoridae): Morphological and Molecular Evidence." *Bulletin of Marine Science* 79, no. 3 (2006): 483–91.

Silvani, L., M. Gazo, and A. Aguilar. "Spanish Driftnet Fishing and Incidental Catches in the Mediterranean." *Biological Conservation* 90, no. 1 (1999): 79–85.

Simões, P. R. "The Swordfish (*Xiphias gladius*) Fishery in the Azores, from 1987 to 1993." *Collective Volume of Scientific Papers ICCAT* 44, no. 3 (1995): 126–31.

Simões, P. R., and J. P. Andrade. "Feeding Dynamics of Swordfish (*Xiphias gladias*) in Azores Area." *Collective Volume of Scientific Papers ICCAT* 51, no. 5 (2000): 1642–46.

Sloan, R. "Raise a Sword." *Saltwater Sportsman* 66, no. 12 (2005): 74–78.

Small, M. F. "The Happy Fat." *New Scientist* 175 (2002): 34–37.

Smith, J. L. B. *High Tide*. [Cape Town]: Books of Africa, 1968.

——. "Pugnacity of Marlins and Swordfish." *Nature* 178 (1956): 1065.

——. *The Sea Fishes of Southern Africa*. [Johannesburg?]: Central News Agency Ltd., 1950.

Smith, J. van C. *Natural History of the Fishes of Massachusetts*. Boston: William D. Ticknor, 1843.

Smith J. R., P. Fong, and R. F. Ambrose. "Dramatic Declines in Mussel Bed Community Diversity: Response to Climate Change." *Ecology* 87, no. 5 (2006): 1153–61.

Smith, M. M., and P. C. Heemstra, eds. *Smiths' Sea Fishes*. Berlin: Springer-Verlag, 1986.

Smith, P. "The Broadbill: The Fish They Called a God." *Sports Afield* 169, no. 4 (1973): 49–50, 177–79.

Sonrel, L. *The Bottom of the Sea*. Trans. and ed. Elihu Rich. New York: Charles Scribner, 1870.

Spiess, A. E. "Who Were They?" *Island Journal* 10 (1992): 24–27.

Spiess, A. E., and R. A. Lewis. *The Turner Farm Fauna: 5000 Years of Hunting and Fishing in Penobscot Bay, Maine*. Augusta: Maine State Museum, Maine Historic Preservation Commission, and Maine Archaeological Society, 2001.

Stanley, S. M. *Extinction*. New York: Scientific American Library, 1987.

——. "Marine Mass Extinctions: A Dominant Role for Temperature." In *Extinctions*, edited by M. H. Nitecki, 69–117. Chicago: University of Chicago Press, 1984.

——. "Mass Extinctions in the Ocean." *Scientific American* 250, no. 6 (1984): 64–72.

Stead, D. G. 1933. *Giants and Pigmies of the Deep*. Sydney: Shakespeare Head Press.

———. 1963. *Sharks and Rays of Australian Seas*. [Sydney]: Angus & Robertson.

Sternberg, C. H. *Hunting Dinosaurs in the Badlands of the Red Deer River, Alberta, Canada*. Lawrence, KS: n.p., 1917. Reprint, Edmonton: Newest Press, 1985.

Stillwell, C. E., and N. E. Kohler. "Food and Feeding Ecology of the Swordfish *Xiphias Gladias* in the Western North Atlantic Ocean with Estimates of Daily Ration." *Marine Ecology Progress Series* 22 (1985): 239–47.

———. "Food, Feeding Habits and Estimates of Daily Ration of the Shortfin Mako (*Isurus oxyrinchus*) in the Northwest Atlantic." *Canadian Journal of Fisheries and Aquatic Sciences* 39 (1982): 407–14.

Stone, H. H., and L. K. Dixon. "A Comparison of Catches of Swordfish, *Xiphias gladius*, and Other Pelagic Species from Canadian Longline Gear Configured with Alternating Monofilament Nylon Gangions." *Fishery Bulletin* 99, no. 1 (2001): 210–16.

Storer, D. H. *Reports on the Fishes, Reptiles and Birds of Massachusetts*. Reports of the Commissioners on the Zoological and Botanical Survey of the State of Massachusetts. Boston: Dutton and Wentworth, 1839.

Stramma, L., G. C. Johnson, J. Sprintall, and V. Mohrholz. "Expanding Oxygen-Minimum Zones in the Tropical Oceans." *Science* 320, no. 5876 (2008): 655–58.

Stroud, R. H., ed. *Planning the Future of Billfishes: Research and Management in the 90s and Beyond. Planning the Future of Billfishes*. Pt. 2, *Contributed Papers*. International Billfish Symposium, Kailua-Kona, Hawaii, August 1–5, 1988. Savannah, GA: National Coalition for Marine Conservation, 1990.

Sugimoto, M. "Morphological Color Changes in Fish: Regulation of Pigment Cell Density and Morphology." *Microscopic Research and Technology* 58, no. 6 (2002): 496–501.

Sun, C.-L., S.-P. Wang, and S.-Z. Yeh. "Age and Growth of the Swordfish (*Xiphias Gladius* L.) in the Waters around Taiwan Determined from Anal-Fin Rays." *Fishery Bulletin* 100, no. 4 (2002): 822–35.

Sytchevskaya, E. K., and A. M. Prokofiev. "First Findings of Xiphioidea (Perciformes) in the Late Paleocene of Turkmenistan." *Journal of Ichthyology* 42 (2002): 227–37.

Taipe, A., C. Yamashiro, L. Mariategui, P. Rojas, and C. Roque. "Distribution and Concentrations of Jumbo Flying Squid (*Dosidicus gigas*) off the Peruvian Coast between 1991 and 1999." *Fisheries Research* 54 (2001): 21–32.

Takahashi, M., H. Okamura, K. Yokawa, and M. Okazaki. "Swimming Behaviour and Migration of a Swordfish Recorded by an Archival Tag." *Marine and Freshwater Research* 54, no. 4 (2003): 527–34.

Talbot, F. H., and J. J. Penrith. "Spearing Behavior in Feeding in the Black Marlin, *Istiompax marlina*." *Copeia* 1962, no. 2 (1962): 168.

Taylor, M. A. "How Tetrapods Feed in Water: A Functional Analysis by Paradigm." *Zoological Journal of the Linnean Society of London* 91 (1987): 171–95

Taylor, R. G., and M. D. Murphy. "Reproductive Biology of the Swordfish *Xiphias gladius* in the Straits of Florida and Adjacent Waters." *Fishery Bulletin* 90, no. 4 (1992): 809–16.

Thomas, C. D., A. Cameron, R. E. Green, M. Bakkenes, L. J. Beaumont, Y. C. Collingham, B. F. N. Erasmus, et al.. "Extinction Risk from Climate Change." *Nature* 427 (2004): 145–48.

Thomas, P. "Environment: Are Squid Vicious?" *Los Angeles Times*, February 10, 2004.

Tibbo, S. N., L. R. Day, and W. F. Doucet. "The Swordfish, (*Xiphias gladius* L.): Its Life History and Economic Importance in the Northwest Atlantic." *Bulletin of the Fisheries Research Board of Canada* 130 (1961): 1–47.

Tinsley, J. B. *The Sailfish: Swashbuckler of the Open Seas*. Gainesville: University of Florida Press, 1984.

Tittensor, D. P., B. Worm, and R. A. Myers. "Macroecological Changes in Exploited Marine Systems." In *Marine Macroecology*, edited by J. D. Witman and K. Roy, 310–37. Chicago: University of Chicago Press, 2006.

Toll, R. B., and S. C. Hess. "Cephalopods in the Diet of the Swordfish, *Xiphias gladius*, from the Florida Straits." *Fishery Bulletin* 79, no. 4 (1981): 765–74.

Torres-Escribiano, S., A. Ruiz, L. Barrios, D. Vélez, and R. Montoro. "Influence of Mercury Bioaccessibility on Exposure Assessment Associated with Consumption of Cooked Predatory Fish in Spain." *Journal of the Science of Food and Agriculture* 91 (2011): 981–86.

Townsend, C. H. "The Swordfish and the Thresher Shark Delusion." *Bulletin of the New York Zoological Society* 26 (1923): 76–80.

———. "Swordfish Taken on Trawl Lines." *Science* 56 (1922): 18–119.

———. "Why Swordfish Strike Ships." *Bulletin of the New York Zoological Society* 27 (1924): 39–40.

Tserpes, G. "Greek Swordfish Fishery." *Collective Volume of Scientific Papers ICCAT* 56, no. 3 (1995): 287–88.

Tserpes, G., P. Peristeraki, and A. Di Natale. "Standardized Catch Rates for Swordfish (*Xiphias Gladius*) from the Italian and Greek Fisheries Operating in the Central-Eastern Mediterranean." *Collective Volume of Scientific Papers ICCAT* 56, no. 3 (2004): 850–59.

Tudela, S., A. K. Kai, F. Maynou, M. El Andalossi, and P. Guglielmi. "Driftnet

Fishing and Biodiversity Conservation: The Case Study of the Large-Scale Moroccan Driftnet Fleet Operating in the Alboran Sea (SW Mediterranean)." *Biological Conservation* 121 (2005): 65–78.

United Nations General Assembly. *Large-Scale Pelagic Driftnet Fishing and Its Impact on the Living Marine Resources of the World's Oceans and Seas.* UNGA Resolution 44/225. December 1989. http://www.un.org/Depts/dhl/resguide/r44.htm.

Uozumi, Y. "Historical Perspective of Global Billfish Stock Assessment." *Marine and Freshwater Research* 54, no. 4 (2003): 555–65.

van der Elst, R., and X. Roxburgh. "Use of the Bill during Feeding in the Black Marlin (*Makaira indica*)." *Copeia* 1981, no. 1 (1981): 215.

Voss, G. "A Contribution to the Life History and Biology of the Sailfish, *Istiophorus americanus* (Cuv. and Val.), in Florida Waters." *Bulletin of Marine Sciences of the Gulf and Caribbean* 3, no. 3 (1953): 206–40.

Voss, G. L. "Hunting Sea Monsters." *Sea Frontiers* 5, no. 3 (1959): 134–46.

———. "Solving the Secrets of the Sailfish." *National Geographic* 54, no. 6 (1956): 858–72.

Waldman, P. "Mercury and Tuna: U.S. Advice Leaves Lots of Questions." *Wall Street Journal*, August 1, 2005, A1–A6.

Walsh, W. "Incidental Catches of Fishes by Hawai'i Longliners." *Pelagic Fisheries Research Program Newsletter* 7, no. 1 (2002): 1–4.

Walters, V., and H. Fierstine. "Measurements of Swimming Speeds of Yellowfin Tuna and Wahoo." *Nature* 202 (1964): 208–9.

Walton, I. *The Compleat Angler*. 1653. Reprint, Mineola, NY: Dover Publications, 2003.

Ward, P. "Swordfish Fisheries and Management Today." *Pelagic Fisheries Research Program Newsletter* 5, no. 4 (2000): 1–6.

Ward, P., and S. Elscot. *Broadbill Swordfish: Status of World Fisheries.* Canberra: Bureau of Rural Sciences, 2000.

Ward, P., J. M. Porter, and S. Elscot. "Broadbill Swordfish: Status of Established Fisheries and Lessons for Developing Fisheries." *Fish and Fisheries* 1, no. 4 (2000): 317.

Ward, R. D., C. A. Reeb, and B. A. Block. *Population structure of Australian swordfish,* Xiphias gladius*: Final Report to the Australian Management Authority, Canberra.* Hobart, Tasmania: CSIRO Marine Research, 2001.

Weart, S. R. *The Discovery of Global Warming*. Cambridge, MA: Harvard University Press, 2003.

Webster, P. J., G. J. Holland, J. A. Curry, and H.-R. Chang. "Changes in Tropical Cyclone Number, Duration, and Intensity in a Warming Environment." *Science* 309 (2005): 1844–46.

White, L. "The Historical Roots of Our Ecologic Crisis." *Science* 155 (1967): 1203–7.

Whitelaw, W. "Recreational Billfish Catches and Gamefishing Facilities of Pacific Island Nations in the Western and Central Pacific Ocean." *Marine and Freshwater Research* 54, no. 4 (2003): 463–71.

Whynott, D. *Giant Bluefin*. New York: North Point Press, 1995.

———. "The Most Expensive Fish in the Sea." *Discover* 20, no. 4 (1999): 80–85.

Wilcox, W. A. "A Man Killed by a Swordfish." *Bulletin of the US Fisheries Commission* 6 (1887): 417.

Wilkens, L. A., M. A. Hofmann, and W. Wojtenek. "The Electrical Sense of the Paddlefish: A Passive System for the Detection and Capture of Zooplanktonic Prey." *Journal of Physiology* 96 (2002): 363–77.

Wood, G. L. *The Guinness Book of Animal Facts and Feats*. Enfield, Middlesex: Guinness Superlatives, 1982.

Wright, K. "The Mother of All Extinctions." *Discover* 22, no. 10 (2001): 28–29.

———. "Our Preferred Poison—Mercury Is Everywhere." *Discover* 26, no. 3 (2005): 58–65.

Wright, P. B. "Blacks and Blues." *Marlin*, November 2007, 54–59.

Wylie, P. *Crunch and Des: Classic Stories of Saltwater Fishing*. New York: Lyons and Burford, 1990.

Yokawa, K., and T. Fukuda. "Recent Status of the Swordfish Catch by the Japanese Longliners in the Atlantic Ocean." *Collective Volume of Scientific Papers ICCAT* 54, no. 5 (2001): 1547–49.

———. "Swordfish Dead Discards and Live Releases by Japanese Longliners in the North Atlantic Ocean in 2000–2002." *Collective Volume of Scientific Papers ICCAT* 56, no. 3 (2004): 967–77.

Young, E. "Why Hot-Eyed Swordfish Always Get Their Squid." *New Scientist* 185, no. 2482 (2005): 18.

Young, J., A. Drake, M. Brickhill, J. Farley, and T. Carter. "Reproductive Dynamics of Broadbill Swordfish, *Xiphias gladius*, in the Domestic Longline Fishery of Eastern Australia." *Marine and Freshwater Research* 54, no. 4 (2003): 315–32.

Young, W. E., and H. Z. Mazet. *Shark! Shark!* New York: Gotham House, 1934.

Zarudzki, E. F. K. "Swordfish Rams the *Alvin*." *Oceanus* 13, no. 4 (1967): 14–18.

Zharov, V. L., N. F. Paliy, V. I. Sauskan, and V. G. Yurov. "Results of Soviet Fisheries Investigations on Atlantic Tuna." *Collective Volume of Scientific Papers ICCAT* 1 (1973): 549–55.

Zimmerman, T. "It's Hard Out Here for a Shrimp." *Outside* 31, no. 7 (2006): 78–85.

Index

Note: A page number in italics refers to an illustration or its caption.